DATE DUE		
FEB 22 '9	MAY 2 7 '97	
APR 30 '96	MAR 2 5 '98	
MAY 27 '96	MAY 18 '00	
MAY 29 '96	JUL 21 '00	
JAN 1 1 '9	OCT 06 '04	
	3/5/05	
JUL 2 96		
SEP 1 3 '96		
OCT 11 '96		
DEC 4 '96		
FE 08 97		

COSMIC VOYAGE

Other Books by Courtney Brown

Ballots of Tumult: A Portrait of Volatility in American Voting

Serpents in the Sand: Essays on the Nonlinear Nature of Politics and Human Destiny

Chaos and Catastrophe Theories

COSMIC VOYAGE

A SCIENTIFIC DISCOVERY OF EXTRATERRESTRIALS VISITING EARTH

COURTNEY BROWN, Ph.D.

A DUTTON BOOK

DUTTON
Published by the Penguin Group
Penguin Books USA Inc., 375 Hudson Street,
New York, New York 10014, U.S.A.
Penguin Books Ltd, 27 Wrights Lane,
London W8 5TZ, England
Penguin Books Australia Ltd, Ringwood,
Victoria, Australia
Penguin Books Canada Ltd, 10 Alcorn Avenue,
Toronto, Ontario, Canada M4V 3B2
Penguin Books (N.Z.) Ltd, 182–190 Wairau Road,
Auckland 10, New Zealand

Penguin Books Ltd, Registered Offices:
Harmondsworth, Middlesex, England

First published by Dutton, an imprint of Dutton Signet, a division
of Penguin Books USA Inc.
Distributed in Canada by McClelland & Stewart Inc.

First Printing, January, 1996
10 9 8 7 6 5 4 3 2 1

REGISTERED TRADEMARK—MARCA REGISTRADA

LIBRARY OF CONGRESS CATALOGING-IN-PUBLICATION DATA:

Brown, Courtney.
    Cosmic voyage : a scientific discovery of extraterrestrials visiting Earth / Courtney Brown.
        p.   cm.
    ISBN 0-525-94098-7
    1. Life on other planets.   2. Extraterrestrial anthropology.
    3. Remote viewing (Parapsychology)   I. Title.
    QB54.B76   1996
    001.9'42—dc20   95–38332
                                                                              CIP

Printed in the United States of America
Set in New Baskerville
Designed by Leonard Telesca

CONTENTS

PART III: A HUMAN APPROACH TO GALACTIC POLITICS

ACKNOWLEDGMENTS

No book such as this one could have been written without the help of many other people. I could not have even begun the research for this book without the assistance of my teachers of the subject of consciousness. I was blind to an important aspect of my own existence before they began to guide my growth in this regard. My concrete awareness of who I am, and who we all are, has matured greatly since that period of near total darkness.

A special note of thanks must be given to my agent and friend, Sandra Martin. She offered to market my book when it was little more than an idea, and she has been supportive of my efforts through all of the difficult and unsure times that preceded the manuscript's completion. She believed in me, and that belief gave me courage to finish a project that has always been almost certain to bring a storm of criticism from many of my academic colleagues.

I am grateful for the support of my editor at Penguin USA, Edward Stackler. He took a chance by supporting publication of this book when the manuscript was splitting the editorial staffs of other presses right down the middle. His emphasis in keeping my writing clear and simple, as compared with a more academic style, is also greatly appreciated. Robert Durant, Jo Lenore Jordan, and Dale Stephens also gave me useful advice.

My wife and son helped in their own ways. It has not been possible for us to live simple, "normal" lives since this all began. Yet through it all, my wife has supported me. As for my son, he enabled me to put my research into a broader context by helping me see the real reason behind all of life's struggles.

There are extraterrestrials, as my readers will see. Make no mistake about it: I could not have gathered any data or written this book if many of these extraterrestrials did not cooperate with my research efforts. Indeed, this is really as much their book as it is mine. It is, at least, their true story that I tell.

Finally, I want to thank in advance all of those readers who will understand and appreciate that which I have done. There are always people ready to criticize novel research, and I will weather their criticism as best I can. The big unknown is the number of people who will find this book useful. However many there are in this latter group, rest assured that I am grateful to you. If your lives are richer in any way because of what I report here, all of my efforts will have been worthwhile.

Prologue

Cosmic Voyage is a detailed examination of two societies of known intelligent extraterrestrial life. More specifically, this volume is the result of years of work observing alien cultures whose activities here on Earth have been very pronounced. The bottom line is that *Cosmic Voyage* describes the history of two alien worlds that died, and how the civilization of each survived beyond its homeworld's death to arrive here, on Earth. These survivors have needs, desperate needs. But as it turns out, so do we humans, and this galactic tryst is leading to a future in which three races share a common destiny. The great link connecting the three races is that all three homeworlds either already experienced, or will soon experience, planet-wide ecological disasters of spectacular proportions. Indeed, it is from these other two races that humans will learn much regarding how others have survived on planets of dust.

The research presented in this book was conducted using rigorous and exacting remote-viewing protocols that were recently developed for the U.S. military for espionage purposes. The data that are obtained using these protocols accurately represent reality, not imagination or allegory. I make no apologies for the methods used to conduct my research, though in the absence of these methods the research would not have been possible. The methods are new, but they are valid and exceptionally reliable research instruments, regardless of whether many other scientists yet accept them or are familiar with them.

What follows in this volume is what I learned about extraterrestrials, both during my own training and in the months that followed. In Part I of this book, I introduce both the subject and the history of remote viewing. Also in Part I, I relate my own story, which is associated with my extensive training program in remote viewing and in other advanced techniques relating to consciousness. The heart of the book is located in Part II, where I detail all of my data and analyses regarding extraterrestrials. I have chosen to present this material in Part II chronologically so that the reader can share with me the thrill of the discoveries as they were made. In Part III of this book, I analyze where humankind now stands, given what we know about our needs and the needs of the broader galactic community. I make suggestions regarding our participation in interspecies diplomacy, including a course of study that our diplomats can follow to begin the representation of humans in the Galactic Federation, a galactic collective organization.

There *is* extraterrestrial life, lots of it. This book explains what we now know about two extraterrestrial civilizations that have recently been visiting Earth. This is not a book of speculation about extraterrestrial life. This is a volume of results, as well as the interpretation of those facts, that I am willing to defend as authoritative regarding existing societies that have evolved on worlds other than our own planet.

There is always a study that is the first of its kind, and this is such a study. Widespread acceptance of the methods will come in time; there is no doubt in my mind about this. Meanwhile, we need not be ashamed of using these newly discovered methods while we wait for a new generation of scientists to become acquainted with them, as long as the uses of the techniques are held to rigorous scientific standards. This book sets a baseline for these standards.

The methods used to collect the data in this book have been as rigorously controlled as those used in any solid social science study. This is not to say that the methods are the same as those typically used in the social sciences, but rather the application of scientific principles in guiding the data-collection procedures have been rigorously followed. As I explain more fully later, this is particularly true of the principles of replicability of results.

Humans have an astonishing ability to dismiss information that does not conform to their preconceived notions of reality. Scientists, being human, suffer from this mind-set as much as anyone else. In some circles, these preconceived worldviews are known as currently accepted paradigms. These are informational patterns, internally constructed templates, with which all externally obtained information is judged. This externally obtained information can come from a newspaper, a friend, a lecturer in a university, a book, or any other source. But when confronted with ideas, let alone facts, that do not fit into an accepted informational paradigm, humans tend to have an intense desire *not* to believe the new information. At times it seems any excuse will seem rational, since the goal is what matters: the established paradigm must not be readily abandoned.

Because of this phenomenon, large numbers of contradictions occur in human society. For example, it is easy to find any number of physicists who will tell you that there is no evidence that supports the idea that telepathy is possible. On the other hand, it is just as easy to determine that many of these same physicists routinely go to places of worship with their families at least once a week to engage in telepathic communication with one or more nonphysical beings. Indeed, as will become clear by the time you finish this book, the vast majority of scientists who do not consider things such as telepathy and remote viewing "real" are simply misinformed at best or, more likely, too biased to look at the subject objectively.

But make no mistake about it. Critical nodes of the conservative scientific community are well aware of the absolute existence of at least some psi phenomena. There are many examples of scientific verification of such phenomena. However, a particularly noteworthy report by two psychologists, Daryl J. Bern and Charles Honorton, on telepathic communication between humans in a series of highly controlled studies appeared in the January 1994 issue of the mainstream psychology journal *Psychological Bulletin*. Certainly large numbers of scientists remain skeptical. But the ultimate outcome of the debate can no longer be doubted. As time marches on, increasing numbers of mainstream scientists will continue to "discover" a wide array of psi phenomena.

The great physicist Max Planck once noted that major advances in the sciences occur not because someone makes an important discovery and everyone else eagerly accepts the new ideas. Rather, generational change mediates the advancement of science. Older scientists tend to stick to the intellectual paradigms that were current during the time when they did the major portion of their research early in their careers. Thus, society often waits until older scientists are replaced by a new generation of scientists who have since the beginning of their careers been acquainted with new ideas.

Over the past fifteen years, scientific understanding of remote viewing—the ability to accurately perceive information at great distances across space and time—has made tremendous advances, though a broad scientific acceptance of this understanding still lags. Throughout human history, it has repeatedly been noted that certain apparently gifted individuals had the ability to perceive information from a remote location—in the sense, for example, that a person could perceive a house on the other side of the planet. But it is precisely because science could not figure out why only "gifted" individuals could do these things (and not always consistently) that the reality of the ability has always been questioned. All that is changed now. The most significant discovery of the past fifteen years is that we do not need to rely on gifted individuals to perform these feats any longer. The talent can be taught, and anyone—including scientists—can learn it and use it with great accuracy. Moreover, the reliability of trained individuals is generally much greater than that of the best natural psychics. Executed competently, studies employing remote viewing using trained viewers can yield replicable results with nearly total accuracy, virtually all of the time.

It was members of the U.S. military who learned this trained form of remote viewing while serving in a highly classified team of special operations and intelligence officers in the Army. The original purpose behind training these psychic warriors was to spy on the perceived enemies of the United States. However, once their training was complete, the group began to view targets that were often more interesting than, say, missile silos or meetings within the Kremlin walls. The group began to examine the enigma of unidentified flying objects and, more specifically, extraterrestrial beings visiting Earth.

My own interaction with this group of military remote viewers began after many of them had left the military in hopes of using these newly developed remote-viewing tools more broadly than had previously been possible. One of my first characterizations of their early efforts with regard to UFOs was that they were concentrating too much of their energy on the beings flying the ships. It was my view that they should shift their efforts entirely to understanding the societies from which the ships emanated. I offered my services as a social scientist to them, hoping that I would be able to make a significant contribution in answering a broader set of questions relating to the structure of sentient life in our galaxy. This was the genesis of this book.

Until now, few have known the complete story of what remote viewing has revealed about the UFO phenomenon. This book represents an attempt to put as many of the pieces of the puzzle together as possible, given our current knowledge. This is not the definitive UFO or extraterrestrial (ET) book. Rather, it is one attempt at solid research using a new set of tools for data gathering. The expectation is that other researchers using these same tools will make further discoveries and that our understanding of ET life will continue to expand.

The Choice of Species

This volume examines the societies and homeworlds of two extraterrestrial civilizations. One, an ancient civilization that flourished on Mars during the time that dinosaurs roamed on Earth, is currently and precariously sustained with a much-diminished population; the other is that of a group of beings called the Greys. The choice of these two civilizations for this volume was not made because remote viewers have found no other civilizations, for we have indeed found others. I focus on these two civilizations because they are playing a particularly important role in the current evolution of our own civilization on this planet.

Certain practical matters also suggest that the Martian and the Grey civilizations are the appropriate targets of investigation at this time. Mars is physically very close to us, and there is a natural interest among humans in that planet's history. Humans will be able to visit Mars in the near future. This will allow us to examine the archaeological ruins of that civilization closely, thereby adding physical evidence to the remote-viewing data that are presented here. With regard to the Greys, they have been closely involved with both Martians and humans for a long time. Given the scope of their activities in this solar system, it only makes sense to explain who the Greys are and to describe their interesting history.

The Stakes

This is a book of fact, not fiction. I have repeatedly checked on the accuracy of my observations under a variety of data-collection settings, and as of mid-1995 other trained remote viewers have independently corroborated many, and perhaps most, of my basic findings. *Thus, replicability is an important element of the claims that I make here.* I am continuing to work with viewers to this day to obtain further corroboration of my results.

Any person of sound mind can now obtain the training that is necessary to independently replicate my results. (I describe this training program in detail in Chapter 35 of this volume.) Replicability is the primary criterion of all *science*. If a discovery is made, the scientist making the discovery needs to explain clearly the procedures that were used to make the discovery. Other scientists then typically duplicate the procedures *exactly* to verify the original claims. *No criticisms of the initial claims are valid in the absence of attempts to duplicate the original experiments.* This is as true for my own research as it is for the research of a physicist who claims to have discovered a new subatomic particle using a particular experimental setup involving a particle accelerator.

What I discovered in the process of my research was more unexpected than the plot of any science fiction novel. I never could have dreamed up a story more amazing than the reality that I have perceived. What I have learned makes sense in retrospect. But learning it challenged nearly every preconception that I had, and I would be lying to my readers if I wrote that the process was an easy one.

I do not publish this material naively. I do not look forward to the almost certain barrage of criticism that will result from the publication of these findings. Moreover, I have an enviable and hard-earned reputation for thoughtful and creative research, often involving sophisticated nonlinear mathematical representations of social phenomena. I do not want to lose this reputation. But a person who is fully committed to science as a profession must accept the responsibility that comes with it. My job as a scientist is not to publish that which is popular or, as the current buzzword goes, "politically correct." A scientist must report the truth, whatever that truth may be, and the potential reaction of others to this truth should never be a primary consideration in the decision to publish competently researched and fully replicable results.

Simply, the human species is at a crossroads in its evolutionary history. We are about to enter the realm of galactic life as fully participating members in the community of worlds. The short-term career considerations of *any* scientist do not weigh significantly against this broader agenda.

This does not mean that others will not modify or enhance my findings. I am not perfect, and future scholars will add to my research and correct what mistakes I have made. But the fundamental aspects of my analyses—I am certain—will be sustained. Those who dare to send their minds where mine has gone will find a truth that needs no human defender.

PART 1

THE PRELIMINARIES

CHAPTER 1

A Brief History of the U.S. Military Psychic Warfare Program

This is a book about two extraterrestrial civilizations that either already have or soon will have an important evolutionary impact on human life on Earth. This is not a book about scientific remote viewing. Nonetheless, since scientific remote viewing has been used to obtain the data presented here, and since it is a new science, it is necessary to briefly outline the history of the subject so that readers can place the techniques used in this research into a proper context.

The U.S. government's involvement with psychic matters was born of the need to collect information about the country's enemies. Initial interest originated within the CIA in the 1970s, but the bulk of the research was conducted by U.S. Army intelligence, beginning with a highly secret project aimed at training some of their best officers.

The critical problem with military intelligence collection has always been the risk faced by its agents, often caused by the difficulty of communicating information back to headquarters. Technological devices—no matter how cleverly they were concealed—could still be discovered, and the agent's life and information jeopardized. What was needed was a means of communicating information to Washington, D.C., without any physical apparatus.

The original idea was for the military to develop some type of psychic switch that could be activated in the Pentagon from, say, Moscow. The agent could be assigned the task of getting some crucial information that would have a yes or no informational response to it. For example, the U.S. military could be interested in whether or not the Soviets had a new type of weapon. A spy tasked with acquiring this information could productively use such a psychic switch. Even if the spy was under surveillance by the KGB, it would never be obvious that he or she was transmitting data.

The U.S. military was also concerned that the Soviets might be developing a psychic military potential. The United States did not want to be left behind, and a psychic cold war developed.

Interestingly, the Soviets actually did have a psychic warfare program. Their approach was to screen their population for the best possible natural psychics rather than to attempt to create a training program. While they did manage to assemble an effective psychic team, the Soviet efforts were hampered by the same problem that plagued the U.S. efforts, namely, resistance from higher levels of command. On both sides, some of the resistance was aimed at preventing a focus on particular unwanted targets, such as UFOs. But sometimes the resistance was much more general to the nature of the data-collection method itself.

Many commanding officers on both the U.S. and Soviet sides subscribe to conservative or traditional belief systems, often religious in nature. Even the nonreligious objections made it clear that many people did not want to acknowledge the capabilities of such techniques. The resistance extended beyond the military. Indeed, I was told of one instance in which a very high-ranking civilian political appointee serving directly under the secretary of defense began to object strenuously during a top secret briefing on the subject of UFOs when the matters of alien technology and psychic information were raised. The official asserted that this information was not supposed to be known by any humans until we died and learned it from heavenly sources. Apparently, the Soviet situation was no better. Their officials were spooked by the subject as well, and their project remained underfunded.

The CIA's initial involvement with psychic matters began by working with natural psychics. When the CIA's covert mining of Nicaraguan harbors came to the attention of Congress, the CIA purged all units and projects that could lead to further political trouble or embarrassment. This ended the CIA's involvement with psychic warfare.

The program to develop a psychic "switch" never succeeded. But the effort to do so resulted in the military's exploration of the application of psychic techniques in intelligence gathering. Two projects are particularly noteworthy in this regard. The first was the work by Professor Robert Jahn at Princeton University's PEAR (Princeton Engineering Anomalies Research) laboratory, which attracted a great deal of interest from the intelligence community, though the PEAR lab received no military or intelligence funding. However, it was the research at the Remote Viewing Lab at SRI International (formerly the Stanford Research Institute) under the leadership of Dr. Harold Puthoff that attracted the military's interest most.

The U.S. Army did not have the same political problems that plagued the CIA. To the Army, mission accomplishment was the only thing that mattered. While the CIA was enmeshed in its psychic troubles, the Army began creating a group of hidden or "black" units that would help solve some of its more difficult intelligence problems.

One of these special units was code-named Detachment G (for "Grill Flame") and did not appear on any organizational schematic for the military. Detachment G was assigned the original task of investigating the use of psychic techniques to penetrate the most secret military projects of the enemies of the United States.

Because of this unit's unusual nature, information it gathered was circulated only to a handful of the highest-ranking officers and political appointees. It soon became apparent that the project was yielding useful information. If the project was going to mature, it would have to be expanded beyond its existing boundaries. The problem with expanding the project was that the phenomenon of remote viewing had not been recognized by the scientific community. The Army needed to find some way to give the phenomenon greater scientific credibility so that it could eventually put its efforts on the books and increase funding. Thus it began funding some scientific efforts in an attempt to validate the phenomenon.

The early efforts at psychic information gathering did not involve remote viewing as it is practiced today. These first efforts focused on maintaining altered states of consciousness in people who were natural psychics. Operationally, this usually involved a psychic lying on a bed with electrodes connected to his or her head and feet. The electronic equipment was used to indicate that the subject had achieved a 180-degree polarity shift in body voltage, which was usually the indication that the altered state had been achieved. Another person in the room, called a "facilitator," would then instruct the person to "move" to the target and report what he or she observed.

Though these experiments yielded some valuable information, the information gathered in this fashion was not always consistent across sessions or subjects. The military needed high degrees of reliability; nothing else would do if the big brass were to be convinced of the material's value.

It was in 1982 that natural psychic Ingo Swann made his major breakthroughs in remote viewing by developing the protocols that would support reliable intelligence gathering. Swann made his discoveries over many years as he was participating in extensive experiments being conducted at various scientific institutes, including Stanford Research Institute (Swann 1991, pp. 92–4). He developed a form of remote viewing that was based on the use of geographical coordinates, and this form became known as "coordinate remote viewing."[1]

Later, Swann was contracted to train over a dozen individuals in these techniques—some members of the military and some civilians. Their original training lasted one year. To introduce the general subject of altered awareness to the trainees, the team was first

---

1. Another natural psychic who has worked extensively with SRI International is Joseph McMoneagle. Mr. McMoneagle has recently published a very readable book on the subject of remote viewing. The book, *Mind Trek: Exploring Consciousness, Time, and Space Through Remote Viewing*, also has a chapter on some of his own viewing of a past civilization on Mars (McMoneagle 1993, pp. 155–74). Indeed, my own research under controlled conditions corroborates many of McMoneagle's observations regarding this ancient Martian civilization.

sent to the Monroe Institute in Virginia, where they received formal training in out-of-body states.

Washington, D.C., would not be itself without at least one major scandal occupying the attention of the lawmakers and the press at any given time. Every so often—but always after a major scandal—the powers-that-be take action to avoid such incidents in the future. During the Iran–Nicaragua–Oliver North fiasco, the secretary of defense initiated a search throughout the defense community for any other rogue or "hip pocket" organizations lacking proper oversight, which might prove politically embarrassing to the president. He found the remote-viewing detachment and sent an inspector general team to investigate it. Since the remote-viewing team was supposed to be a research unit, the civilian overseers presented the research they thought could be defended as "normal."

Operational matters went from bad to worse from that point on for the nation's most highly trained remote viewers. Their influence in Washington, never great, diminished.

Yet all of the team members knew by this time that they had been given a special gift, a gift of sight. This gift brought with it a responsibility that extended beyond national boundaries. It was this realization, together with a parallel and newly born need to serve a greater cause, that beckoned some of them to turn their inner eye upward, toward the stars. When this all began in the early 1980s, none of them ever could have guessed that their gaze would eventually lead to a mission that could alter the evolutionary course of humanity itself.

CHAPTER 2

Remote Viewing

Background

Probably the best single course of information regarding the historical origins of modern remote viewing (i.e., the protocols themselves, not the military program to use them) is a book written by the person who developed the early version of the actual remote-viewing protocols used by the U.S. military, Ingo Swann. In his book *Everybody's Guide to Natural ESP*, Swann describes a basic theoretical overview of why remote viewing works (Swann 1991). It should be understood that Swann's views are hypotheses, or theories, regarding remote viewing. Swann is an artist (a painter) and an extraordinarily gifted natural psychic, but he is not a scientist. Nonetheless, his views are a valuable set of intuitively guided ideas on the matter.

Readers should understand at the outset that remote viewing (as the term is used here) bears no similarity to the techniques of television or tabloid psychics. Remote viewing is an exacting and demanding discipline that involves a precisely structured set of protocols, and only an individual who has been fully trained by a competent teacher can utilize it accurately for data-gathering purposes. Readers of this book would be well advised to put aside—at least temporarily—any opinions (pro or con) that may be based on previously obtained information or experience with natural psychics. Both methodologically and substantively, this book contains information that will be totally new to nearly all readers.

Scientific Remote Viewing

Remote viewing has evolved from an art to a science through a striking history of progress and refinement, and the use of remote viewing in the course of my research also led to enhancements in both technique and understanding. In the 1980s, the primary innovation of the modern military procedures over coordinate remote viewing was that the restrictive need to use geographical coordinates was eliminated. But also, the modern military version of remote viewing can gather greater quantities and different qualities of data than traditionally was the case with coordinate remote viewing.

Interestingly, private companies now exist that use these military-developed procedures. Moreover, the procedures are often variously labeled, depending on who is using them. Even my own trainer in the military-developed procedures has renamed them. But to my knowledge, these variously labeled procedures are identical, or nearly identical, to those which were developed by the U.S. military, and which are still in use by the military today.

The form of remote viewing used to conduct the research for this book I now call "scientified remote viewing." Scientific remote viewing (SRV) is a technique that is derived from the military-developed procedures as well. Scientific remote viewing is slightly different from the modern military procedures because of how, and for what purpose, it is used. SRV is identical to the modern military procedures for remote viewing in terms of structure. But SRV has been extended to enable two-way communication between a remote viewer and telepathically capable beings. The military version of remote viewing was always a passive data acquisition procedure, and it was never used for communication purposes in this fashion. Nonetheless, in the discussion below, I describe the structure of the military-derived remote-viewing procedures, not their use. Thus, when I refer to the structure of SRV,

I am referring more generally to the modern military-developed procedures as well.

SRV is a set of procedures, or procedures, that allows what is often referred to as the "unconscious mind" to communicate with the conscious mind, thereby transferring valuable information from one level of awareness to another. Information coming from the unconscious mind is typically considered intuition. It is a feeling about something of which one otherwise has no direct knowledge. For example, many mothers will claim that they simply know when one of their children is in serious trouble. They feel it in their bones, so to speak, even when they have not been told anything specific regarding their child's situation. More generally, intuition operates across space and time without any physical means of information transference. SRV systematizes the reading of intuition and allows it to be *accurately* transcribed onto paper, and later analyzed.

Using scientific remote viewing, the information coming from the unconscious is recorded before the conscious mind has a chance to interfere with it using normal waking-state intellectual processes, such as rationalization or imagination. Parts of these protocols are very similar to the picture drawings of remote objects that have been described in the extant historical literature on the subject. (See Swann 1991, pp. 73–114, for a useful review of this literature.) Indeed, picture drawings are a crucial component of the first and third stages of SRV, and remote viewers are trained to decode these drawings in order to extract basic information about the target (i.e., that which is being remote-viewed).

Basically, information about a target comes to trained individuals through their unconscious minds. Remote viewers quickly write down this information during a remote-viewing session while staying within the strict structure of the protocols. The rules of SRV enable a viewer to avoid using the intellectual processes of his or her conscious mind until after the remote-viewing session is completed. Deviating from the protocols even slightly invites the conscious mind to intervene in the process. To do this would court disaster, since the conscious mind would try to interpret data on the spot, thereby activating the mind's imagination. Experience has shown that this seriously compromises the accuracy of the data, which is why untrained natural psychics are generally not reliable remote viewers. Not analyzing the data until after it is collected is the single most important characteristic of SRV. Without this, remote viewing is no more reliable than having a daytime fantasy.[2]

The following point is extremely important. I am not asking anyone to believe what I write in this book, in the sense that one must believe a set of religious ideas. This book is a report of my investigations. As with all good scientific investigations, this one is independently replicable by anyone trained in the protocols of SRV. Thus, other researchers can corroborate everything that I report here. Moreover, I have already gone to great lengths to document and corroborate all of the information reported here. While the mechanics of this corroboration are described later in this chapter, it is important to emphasize at this point that faith or belief has no place in this or any other scientific investigation. Only data and the intelligent interpretation of these data matter. Here, I present and interpret a large body of data. Other researchers can verify the accuracy of these data easily, as long as they are appropriately trained in the protocols of SRV.

Target Coordinates

Scientific remote viewing always focuses on a target. A target can be almost anything about which one needs information. Typically, targets are places, events, or people. But one can also work with more challenging targets as well, such as a person's dreams, or even God. One relies on the unconscious to deliver the required information in a way that will be understandable to the conscious mind.

An SRV session begins by executing a set of procedures using target coordinates. These are essentially two randomly generated four-digit numbers that are assigned to the target, and the remote viewer does not have to know what target the numbers represent. It is convenient to use numbers for these coordinates, but letters would work as well. These coordinates are, obviously, not indica-

2. For additional general information on the role of the unconscious in information transference, see Targ and Puthoff 1977, Wilber 1977, and Mavromatis 1987.

tive of a target's geographic location. The numbers are themselves meaningless to the conscious mind of the remote viewer.

Using these numbers rather than, say, the name of the target, helps distance the conscious mind and its imagination from the data-collection process, thereby inhibiting guessing and other forms of data contamination. Moreover, extensive experience has demonstrated that the unconscious mind instantly knows the target even if it is only given its coordinate numbers. Indeed, in practice, target coordinates are often given to a remote viewer without any additional information. The remote viewer then conducts the SRV protocols on these numbers to obtain target information without being told the target's identity until after the remote viewing is completed.

Using these target coordinates, a remote viewer would follow the strict protocols of SRV throughout the session. The mental connection with the target produces what is called a *signal*. All information coming from the target is distinguished from contaminating information (such as from the imagination) by the viewer's learning to discern the distinct mental flavor of signal information. At the end of each session, the viewer is given the actual description of the target to allow a comparison with the remote-viewing data, thereby obtaining feedback on the data-gathering process.

The SRV Protocols

The SRV protocols have seven distinct stages. In each stage, different types of information are obtained from the target. The stages are engaged in an SRV session sequentially, from Stage 1 to Stage 7, although often a session will end without completing all seven stages if the needed target information has been obtained using the earlier stages only.

The seven stages of the SRV protocols are as follows:

- *Stage 1:* Stages 1 and 2 are referred to as "the preliminaries" in this book and are designed to establish initial site contact. The data obtained about the target in Stage 1—for example, whether there is a man-made structure associated with the target site—are crude.
- *Stage 2:* This stage increases the contact with the site. Infor-

mation obtained in this stage includes colors, surface textures, temperatures, tastes, smells, and sounds that are associated with the target.

- *Stage 3:* This stage involves an initial sketch of the target.
- *Stage 4:* Target contact in this stage is quite intimate. In Stage 4, the unconscious is allowed total control in "solving the problem" by permitting it to direct the flow of information to the conscious mind.
- *Stage 5:* This stage obtains details regarding particular structures, such as the furniture in a room. This stage is often omitted in SRV sessions unless such detailed information regarding a particular object is required.
- *Stage 6:* In this stage, the remote viewer can conduct some guided explorations of the site. The viewer can engage in some limited conscious intellectual activity to direct the unconscious to do certain specific tasks. This is where timelines and geographic locational arrangements are analyzed. Advanced sketches are also drawn in this stage.
- *Stage 7:* This stage is used to obtain auditory information relating to the site, such as the name of a location.

Categories of Remote-Viewing Data

Not all remote-viewing data are the same. Indeed, there are various types of data, all obtained under very different conditions. Remote viewing, under any conditions, is not easy to do. One does not close one's eyes and suddenly "see" the target. The process takes approximately one hour per session, and multiple sessions per target are often needed in order to get a firm grasp of the objects, beings, ideas, and so forth that are associated with the target.

There are six different types of remote-viewing data. One distinguishing characteristic of the various types of data is the amount of information the viewer has about the target prior to the beginning of the remote-viewing session. This information often differs from session to session. The other primary distinguishing characteristic of the data types is whether or not the viewer is working with a person called a monitor, as I explain more thoroughly below.

Depending on the purpose of the session, there can be, say, six hundred separate things to do—one quickly following another—within up to seven distinct stages of the protocols. The basic idea behind these many tasks is to record (on paper) target information as quickly as possible before the analytic portions of the mind can distort, interpret, or otherwise contaminate it. At the end of a session, the viewer has approximately twenty sheets of paper with various forms of data, which are then decoded, interpreted, and summarized.

Target contact during SRV can sometimes be intimate. It often happens that approximately halfway through the session, the viewer begins to experience bilocation, in which the viewer feels he or she is at two places at once. The rate at which data come through from the remote-viewing signal at this point is often very fast, and it is necessary for the viewer to record as much as possible in a relatively short period of time.

Type 1 data

When a remote viewer conducts a session alone, the conditions of data collection are referred to as *solo*. When the session is solo and the remote viewer picks the target (thus having prior knowledge of that target), the data are referenced as *Type 1* data.

Knowing the target in advance is called *front-loading*. Front-loading is often necessary; sometimes a viewer simply needs to know something about a known target. The difficulty with this type of session (primarily affecting novice remote viewers) is that the viewer's imagination can more easily contaminate these data, since the viewer may have preconceived notions of the target. This is why it is so important to follow the exact structure of the remote-viewing protocols and thereby limit this type of contamination. The risk of contamination diminishes markedly with experience, which cultivates a habit of staying strictly within the structure of the protocols.

Type 2 data

For the novice remote viewer, the risk of contamination is reduced with *Type 2* data. With this type of remote-viewing session, the viewer works solo but does not choose the target for the particular session. The target is selected by a computer at random from a predetermined list of targets; the computer supplies the viewer with only the coordinates for the target. The viewer may be familiar with the list of targets (and, indeed, may have been involved in choosing the targets for the list), but only the computer knows which numbers are associated with each specific target. Since the conscious minds of the remote viewers do not know which target is associated with which coordinates, viewers must use their unconscious to extract all information regarding the target. Thus, it is said that the viewer is conducting the session *blind,* which means without prior knowledge (or front-loading) of the target.

Type 3 data

Another type of solo and blind session is used to collect what is called *Type 3* data. With Type 3 data, the target is determined by someone other than the remote viewer. For example, a remote-viewing company can send a fax transmission from its headquarters containing the target coordinates to a group of trained remote viewers who live across the United States. The company's management knows the target, but the viewers do not. The viewers do not have contact with one another. They may also receive some limited and uncompromising information regarding the target—perhaps whether the target is a place or an event. These viewers then conduct sessions using the coordinates alone and then fax their results back to company headquarters. Experience has shown that information which is corroborated using multiple viewers tends to be accurate 100 percent of the time. Moreover, since the viewers may "drop into" a target at different points in time or space, the different sessions can reveal complementary perspectives of the target, resulting in a more complete picture.

Remote viewing solo does have some drawbacks. When viewers conduct their own SRV sessions, the protocols prevent them from fully using the analytic portions of their minds. Thus, the viewers can find themselves viewing a target without knowing what to do next. Solo sessions yield valuable information about a target, but more detailed and in-depth information can be obtained when someone else is doing the navigation. This other person is called a *monitor*, and monitored sessions can be spectacularly interesting events.

Type 4 data

There are three types of *monitored* SRV sessions. When the monitor knows the target but communicates only the target's coordinates to the viewer, this generates *Type 4* data. These types of monitored sessions are sometimes used heavily during training. Type 4 data can be very useful from a research perspective, since the monitor has the maximum amount of information with which to direct the viewer. In these sessions, the monitor tells the viewer what to do, where to look, where to go, and even what to ask if a telepathic being is encountered. This allows the viewer to almost totally disengage his or her analytic mental abilities while the monitor does all of the analysis.

The monitor and the remote viewer need not be in the same room during a session. Speakerphones can be used to establish the necessary verbal dialogue between the monitor and the viewer. This allows monitored sessions to take place even though the monitor and the remote viewer may be in different locations separated by thousands of miles. Once or twice during such sessions, diagrammatic data can be faxed to the monitor to ensure adequate control of the flow of information. Such situations are referred to as *remotely monitored sessions.* Much of the primary data used for this book came from Type 4 data of this sort.

Type 5 data

In particularly critical situations, researchers may want a totally blind setting for data collection, thereby eliminating any possibility of monitor leading. In these cases, both the viewer and the monitor are blind, with the target's coordinates coming either from an outside agency or drawn by a computer program from a list of targets.[3] Data collected in this manner are called

3. Again, it does not matter if the monitor and the viewer are aware of the contents of the list if the list is long. Experience has shown that if the list is sufficiently long, the conscious mind abandons any attempts to guess the target identity.

Type 5 data. Sessions conducted under these conditions tend to be highly reliable. The disadvantages are that such sessions consume more time than other types of remote viewing, and they may not allow the monitor to sort out the most useful information during the session. It is a bit like asking a flight navigator to begin his or her job after a plane is under way. The flight will probably go more smoothly if a general flight plan is arranged prior to departure. Nonetheless, Type 5 data are extremely useful in some situations, and it can add an extra layer of reliability to the overall results.

Type 6 data

The final sort, *Type 6* data, come from sessions in which both the monitor and the viewer are front-loaded with target information. This type of session is used if the viewer needs to obtain more information about a specific target but feels constrained with solo sessions. In this setting, the monitor takes over the navigation, but the viewer and the monitor communicate in advance with regard to the goals of the session.

Summary of data types

In summary, the different categories of remote-viewing data are

- Type 1: Solo, front-loaded, with target selected by viewer
- Type 2: Solo, blind, with target selected at random by computer from a predetermined list of targets
- Type 3: Solo, blind, with target determined by an outside agency
- Type 4: Monitored, viewer blind and monitor front-loaded
- Type 5: Monitored, viewer and monitor blind, with target selected at random by computer from a predetermined list of targets or by an outside agency
- Type 6: Monitored, viewer and monitor front-loaded

No one type of datum is better than all of the others, and each has its advantages and disadvantages.

List of Targets

The vast majority of the SRV sessions that were used to generate the data for this book were conducted under Type 4 conditions. This means that I, the remote viewer, did the session blind while my monitor knew the target. Most of the targets for the sessions were drawn randomly from a list of forty targets that my monitor and I compiled together. Over the course of my investigations, I allowed my monitor to add approximately fifteen other targets without telling me what they were. My monitor also gave me no advance warning as to whether a target I was about to view was one of these special targets or one from the original list.

After my monitor and I began conducting the sessions, I purposely refused to look at the list of the original forty targets: I wished to avoid keeping track of our (nonsequential) progress through the list. Thus, to say that the data were collected under blind conditions does not mean that I never saw the original forty-item target list. It means that I had no knowledge of which target I was getting on a session-by-session basis.

From the perspective of the SRV protocols, the goal is simply to convince the conscious mind in advance that it is hopeless to attempt to guess target information. This forces the conscious mind to rely totally on the data that are supplied by the unconscious mind. Such measures are not so important for the experienced remote viewer, who is highly skilled in reporting only information that is supplied by the unconscious regardless of the type of data. But given the controversial nature of the subject of this book, I made the decision early in these investigations to add the extra layer of credibility to my data-collection efforts by relying on Type 4 data as much as possible.

Criticisms Born of Ignorance

Critics of remote viewing usually focus on monitored data, and they typically contend that such data (Types 4 through 6) can be contaminated by the monitor's own prejudices and interpretations. Specifically, they charge that the monitor leads the viewer, much in the same way that hand movements have been employed by some therapists, either knowingly or unknowingly, to lead the communications of autistic children. Such criticisms are most often used to discount the validity of remote-viewing data in general. However, these charges against SRV are really quite hollow.

One must remember that most remote-viewing data using Earth-based targets can be independently confirmed, and this has been done exhaustively in the development of these protocols. Thus, we are not dealing with something that requires one to *believe* the data, as one would believe a set of religious ideas. The data are simply accurate or they are not accurate. If the target information cannot be confirmed through physical means, it is always possible for any number of other remote viewers to view the target under solo and blind conditions (Type 3 data) to obtain corroborating data. The probability of multiple remote viewers obtaining the same site information under blind Type 3 conditions is infinitesimal, and it far exceeds the statistical requirements normally imposed on rigorously conducted, empirical scientific research.

These problems of criticism are typical of those experienced by most early explorers. For example, James Bruce was one of the first European explorers to enter the area of Africa that is now known as Ethiopia. His explorations took place in the latter half of the 1700s, and included a large number of experiences that his contemporaries in Britain simply did not believe. People called him liar; they claimed that no place could be that strange. Yet Ethiopia was still Ethiopia, and later explorers found exactly what James Bruce found (in particular, see Hibbert 1982, pp. 21–52). Calling Bruce a liar made little sense, yet people still did it even though they had no personal experience to match his own. Corroborating his experiences was the only thing that made sense.

Disbelief by those who have never had direct experience with exploration is a common human phenomenon that is still with us today. The answer to any doubts regarding a person's experiences is never argumentation about what is or is not possible, but rather corroboration.

Limitations of SRV

Scientific remote viewing has its limitations. Some of these limitations seem to be due to the nature of particular targets. For example, a trained remote viewer can target a book and get a basic idea of its contents, but it may prove impossible to *read* it. I have personally remote-viewed an insignia on the uniform of an ET. I could tell that the uniform was white, but I had to spend a considerable amount of time making out the exact outline of the symbol on the badge. It is similarly difficult to read road signs and street names, though these things can be done by a viewer with a significant level of training.

Another limitation with remote viewing is determining one's location relative to some known position. For example, it is easy to target the homeworld of an ET civilization, but difficult to figure out where that homeworld is in relation to our own solar system. A remote viewer can follow an ET ship from Earth to the ET's homeworld and not know the exact path taken. This type of limitation can be overcome, but the "price" of the information (in terms of time, effort, and resources) is great. To use another example, in the course of our research, we determined that one group of ETs had an underground base on Earth, situated beneath a rounded mountain. With the combined efforts of many remote viewers, we finally determined the rounded mountain's likely location, but it was not an easy task.

Most of SRV's limitations can be overcome with adequate time, effort, and resources. But until very recently, there has been another type of limitation that is of an entirely different nature. Sometimes a remote viewer can be prohibited by an external source from viewing at target. For example, the UFO abductee literature has numerous references to ETs that are known as "Greys." These ETs are short, thin, and greyish in complexion and are often reported making medical examinations and performing gynecological procedures on humans taken aboard spacecraft. Trained remote viewers who have tried to penetrate the Greys' ships have found their vision "blocked." Actually, it would be more accurate to say that a substitute view has been given to the remote viewers.

It is usually easy to detect a fraudulent signal. For example, when the Greys generate this substitute view, multiple remote viewers will receive little if any corroborating data; nothing overlaps. (Recently, as will become clear in Chapter 14, even the prohibition on remote-viewing UFO abductions has been removed.)

A final limitation worth mentioning is one of concept. Remote viewers go into a session "as they are." Remote viewing is a lot like being blindfolded and dropped into a foreign city. You take off the blindfold and look around. You have no idea of where you are, yet you notice buildings, people, strange languages, and many physical sensations. You may be able to perceive everything, but you may not understand anything.

To be understood, all remote-viewing data needs to be placed somewhere within the viewer's own intellectual background. While the unconscious mind tries to make the information understandable to the conscious mind, the job is easier if the conscious mind already understands basic *concepts* related to the viewed data. For example, I would not be very useful remote-viewing details of an advanced alien technology. I simply would not know what I was looking at. Try as it may, my unconscious mind would not likely be able to get my conscious mind to understand anything other than the most basic information regarding the technology. But a trained engineer might be able to grasp all sorts of important information, including technical details. The engineer's education helps with the understanding of what is perceived. On the other hand, since I am a social scientist, I can do a very good job remote-viewing ET societies, and I can understand how they organize and govern themselves. That is, my conscious mind can understand what my unconscious mind has shown me, and I can explain what I see to others. In short, the unconscious mind can perceive virtually anything, but one still relies on the conscious mind to understand what is perceived.

In general, remote viewing's limitations are relatively few, and insofar as these obstacles are a consequence of our skills or training, they will fall before future generations of better-trained remote viewers. As for limitations imposed on us by the ETs, the ETs rarely do this, and when they do, their purpose tends to be to prevent us from interfering with ET activities, or to protect us from something for which we are not fully prepared. I believe that even these limitations may be overcome as humankind matures sufficiently to be introduced to what we now know is a rather robust galactic society.

Biases in the Extant UFO Literature

When compared with what we now know about extraterrestrial activities on or near Earth, the extant literature on UFOs has many scientifically unsupportable biases. However, many of the faulty conclusions that one finds in the literature are not due to incompetence. While conducting the research presented in this book, I have both spoken to and read many books by intelligent people who are sincerely trying to unravel an extremely perplexing problem. The UFO enigma is difficult to understand even with remote-viewing data. In the absence of these data, it is almost impossible to fathom. The only other publicly available sources of information are based on either eyewitness accounts of passing UFOs, abduction reports—usually extracted from hypnotized individuals—or channeled information in which friendly ETs purportedly speak to supportive humans while the humans are in a trancelike state.

There are problems with all of these latter types of data, and it is important to explain these problems clearly. To begin with eyewitness reports of passing UFOs, these reports simply do not contain a sufficient amount of information to be useful from a scientific perspective. What can one say about such a report except that an unusual flying object was sighted? We still are left with no information regarding the occupants of the craft, nor do we know anything about the societies from which they originate.

Problems associated with abduction reports are much more complicated and require a more extended treatment. I have no doubt that something very real is happening to many people who claim to have been abducted by ETs and brought aboard a UFO, and we now have remote-viewing data corroborating much of what has been reported.

The general idea in these reports is that humans are being abducted and used against their will to act as incubators during the first three months of pregnancy for genetically engineered offspring that are part ET and part human. The literature tends to suggest that the Greys have a need to produce a new type of body for themselves, in the sense that they are unsatisfied with their current biology. The vast majority of the abduction cases involve some degree of amnesia on the part of the abductee that is overcome using hypnosis by a trained therapist. Two systematic and useful reports of this phenomenon presented by academics can be found in *Abduction* by Harvard University professor John E. Mack (1994) and *Secret Life: Firsthand Documented Accounts of UFO Abductions* by Temple University professor David Jacobs (1992).

Dr. Mack, a professor of psychiatry at Harvard Medical School, has suggested—based on population surveys—that more than one million Americans may have been abducted at least once during their lives (Jacobs 1992, p. 9), and such experiences are not limited to people of the United States. Some individuals seek help—usually in the form of psychological counseling—to deal with the emotional impact, though the number of these individuals is relatively small. Even so, I have been told that informal reports from some therapists suggest that approximately forty thousand individuals in the United States (to date) may have sought some form of professional help with regard to their abduction experiences. I do not know if this is an accurate estimate.

It is quite possible, perhaps likely, that the people who do go to counselors are primarily those who are particularly disturbed by their encounters with the ETs.[4] This could be because some of the ET-human interactions do not go smoothly, whereas most other interactions are not as traumatic—for whatever reasons. If this is the case, then the abduction literature is biased in terms of its sample of abductees. This is the first of five major biases that I find in this literature. That is, the literature is not working with a representative sample of individuals who have been abducted. On the contrary, using such a potentially unrepresentative sample would skew the results of this research in the direction of interpretations filled with fear and trauma. Predictably, much of this literature (notably excluding work by Mack) is filled with warnings with regard to evil aliens that are abducting adults and snatching fetuses for Nazi-like

---

4. This point has also been raised by Whitley Strieber (1995) in his book *Breakthrough: The Next Step*.

genetic experiments. The point is not whether this is true, but that it is impossible to determine the true character of the ETs based only on a lopsided selection of traumatized individuals. A more balanced selection of cases would include those in which the human's interactions with the ETs went smoothly. But such individuals would not be known to the therapists, since they would probably not seek counseling.

The second major bias involved in nearly all of the abduction literature revolves around the use of hypnosis. I have no doubt that hypnosis has been very helpful in resurrecting memories for abductees. However, if the Greys have the ability to influence memory recall as profoundly as is suggested in this literature, then it is probable that the memories that are recalled are also unrepresentative. Moreover, the memories that would most likely surface are those that contain the greatest emotional impact for the abductee, such as those associated with moments of shock or trauma.

Thus, we have a situation in which an unrepresentative sample of memories may be resurrected from the minds of an unrepresentative sample of abductees. It is not possible to draw general conclusions with regard to overall ET intentions from such data, even if the reported data are accurate. It would be comparable to ETs trying to find out what humans are like by interviewing only automobile accident victims. The results of any such study would inevitably be that humans are a careless, often drunk, sadistic, and evil species that enjoys inflicting suffering on their own kind. Such suffering may occur; but the characterization of that suffering as representative of the entirety of human culture would be terribly misleading.

The third major bias in the abduction literature has to do with the perspective of the researchers themselves. It is easy to sympathize with individuals who feel they were kidnapped and abused. We are a highly emotional species that easily identifies with victims of traumatic events. We hate the villains and seek retribution. Researchers are in the emotional soup—so to speak—with the victims when they conduct hypnosis sessions that bring back hidden memories, and few of these researchers are trained (Mack, notably excluded) to distance themselves emotionally from the lives of their subjects. Only professionals highly trained in the discipline of psychological counseling are likely to be able to work competently with such repressed memories and still remain reasonably objective. Emotions are real, and they must be worked with in the most controlled and competent environments. Minimally, the health of the patients requires this. But also, from an interpretive perspective, to draw general conclusions from data gathered by individuals who are not well trained in psychotherapeutic skills only invites serious misunderstandings of the data.

The fourth bias that I find in the abduction literature involves the broader culture from within which the reports originate. We are a society that loves to report violence. This is clearly evident in most of our daily television news shows. In such programs, one almost never hears about calm and emotionally healthful events. Rather, the news that dominates the airwaves is about rapes, murders, and crimes of all sorts. Victims are typically portrayed in a pathetic fashion. Rarely do news reports say anything sympathetic about a person who commits a crime, like how the person became psychologically unstable when he was sexually abused as a child or a victim of gang rape in an alley of an inner-city slum. We rarely ask *why* a crime has occurred. Rather, we ask how we can punish those who commit crimes, immaturely refusing to develop a more balanced perspective of life's complexities.

Additionally, our culture enjoys dwelling on the details of both imaginary and real violence. Violence is one of the most successful products of the Hollywood film industry. It permeates an enormous number of box office hits. If we ever mature as a society, one of the things that we will have to face is our love of violence. This is not to say that the abductees have not experienced things that they authentically see as personal abuse. But there may be another side to the story that we miss if we only focus on, and amplify, the reports of abuse while never asking if there might be a good reason for what happened. To use an analogy, just as children inevitably feel that they are assaulted when they are taken to the doctor's office to receive their vaccinations, we humans may also feel that we are abused by ETs if we do not understand the broader picture within which the activity takes place. Again, I am not minimizing the experiences of the abductees. I am just arguing for a pause in the storm of fear and hate to develop a more balanced perspective of events before we jump to conclusions and decide that we are under attack.

The final bias that I find in the abduction literature is likely to be the most controversial, and some may react strongly to this idea. The bias is one of racial stereotyping. It is important to understand that I am *not* saying that the abductees are racist. The problem is one of how our general society tries to view beings who are different from ourselves.

According to much of the abduction literature, the ETs who are involved in this activity are quite literally grey. They are small, thin, have large wraparound eyes, leathery skin, and lack emotional depth. They are not tall, blond, and blue-eyed. In my view, these perceived differences have triggered an automatic stereotyping to occur within our own minds with respect to these beings. If we were a society in which racism was absent, I would not raise the point here. But if we are to be honest with ourselves, we must admit that human society has a habitual problem of unfairly establishing rules of human treatment based on racial characteristics. If this is true with how we treat other humans of different races, how more so would it likely be with regard to nonhumans?

Abductees come from our own culture, and our own culture does not always view beings who look significantly different from ourselves in a positive light. Given such a setting, is it not surprising that such beings are often portrayed as evil? From this perspective, the potential exists for humans to contaminate the data with interpretive cultural biases. If we are to play a mature role in galactic society, we will probably need to face our own psychological problems squarely, and we will have to learn how to view other cultures through a more objective lens.

Let me give a rather typical example of the types of stereotyping that appear repeatedly in the abduction literature that points to racism as a serious problem with our view of the universe. With this example, I do not mean to single out critically one author. The example is not unique, and it is used here only for heuristic purposes to illustrate the more general problem. This example comes from a recently published book by George C. Andrews. He relays information given to him from an abductee, and he writes that other abductees have supported many of these views as well. In this book, he presents claims that the Greys were in contact with Hitler, derive nourishment from glandular secretions extracted from mutilated animals, worked with the CIA and the Nazis to deploy the AIDS virus and other viruses, and have done many other negative things. On the other hand, their enemies (i.e., the ET good guys) are "tall Blonds" or "Swedes." These beings are usually beautiful and handsome from a mainstream Western cultural perspective. The Blonds are upset with humans because our governments seem to want to work with the Greys, their mortal and evil enemy (Andrews 1993, pp. 141–64).

These types of racial biases and stereotypical generalizations are particularly unhelpful if we wish to understand the ET phenomenon clearly. Again, it needs repeating: the only things that matter are obtaining accurate data and the intelligent interpretation of these data. Until now, we have had mostly biased data, the interpretation of which has been seriously skewed by our research practices and our own cultural problems. We need a fresh crack at the ET enigma. We need to leave our prejudices behind and to develop a more complete picture of the phenomenon.

The final method of data gathering that I mentioned above is channeling, and some may feel that this is an alternative to remote viewing. It is important to explain here why this is not the case. Channeling occurs when a person goes into a trancelike state while he or she communicates telepathically with an alien entity. Sometimes it is claimed that the entity temporarily "takes over" the person's mind and body in order to make the communication.

In my experience, with few exceptions, I have not found channeled information to be reliably accurate when compared with remote-viewing data. Typically, channelers introduce humans to ET brothers and sisters who say that they are the good guys and who warn humans to watch out for the bad guys. The channeled beings then proceed to offer a variety of information regarding the past, present, and future, together with warnings and admonitions. But the material typically is not consistent across channelers, nor (again) does it usually agree with remote-viewing results. With remote viewing, researchers can control both what is observed, the training of the observer, and the conditions under which the data are observed. Inaccurate data can be isolated and discarded using rigorous screening and checking procedures. But channeling allows none of this. One simply has to believe the channeler or not, and beliefs are not a satisfying substitute for objective observations that can be verified and corroborated independently.

CHAPTER 3

Journey through Akasha

My work in the remote viewing of extraterrestrials began in January of 1992, although at the time I did not know this. Since that date, I have had extensive experience with a variety of extraterrestrials. Most of this contact has been through remote viewing. However, I do not believe that I independently approached the ETs without their awareness. Indeed, I strongly suspect (although I have no proof of this) that suggestions may have been given to me without my awareness to explore the general concept of consciousness as a precursor to what at least some ETs knew would be an active interest in non-Earth civilizations. I am not sure who may have given me these suggestions, but my intuitions clearly suggest (and a good deal of circumstantial evidence supports the idea) that I have been unknowingly guided, or prodded, in this direction.

My explorations into human consciousness involved three stages of training. It began with my learning an advanced version of Transcendental Meditation called the Sidhis. The second stage of training was a week-long course at the Monroe Institute in Faber, Virginia. The third stage was training in remote viewing. Each stage had direct relevance to my ET research, as I explain below.

The Sidhis

In 1991, I began training in the TM-Sidhi Program. Like Transcendental Meditation (TM), the Sidhis are taught by teachers trained by Maharishi Mahesh Yogi. I discuss my experiences with the TM-Sidhi Program for two reasons. First, during my research of extraterrestrial beings, I have encountered groups of extraterrestrials who apparently practice something like the Sidhis themselves; second, during my remote-viewing sessions I have repeatedly experienced the consciousness of one group of extraterrestrials that has a collective mentality. This ET collective consciousness—one may call it a "mass mind"—has a subjective experience similar to that which humans experience during the practice of the TM-Sidhi Program.

My original interest in the TM-Sidhi Program corresponded with the publication of an unusual report in a prestigious social science journal. In the December 1988 issue of the *Journal of Conflict Resolution*, a methodologically sophisticated article appeared which claimed that groups of meditators practicing Transcendental Meditation and the more involved TM-Sidhi Program in one place could influence the level of conflict in nearby locations (Orme-Johnson et al. 1988). This phenomenon is labeled the "Maharishi Effect" in honor of Maharishi Mahesh Yogi. The article was considered controversial when it was published, and, by all accounts, it still is.

Following the appearance of the article on the Maharishi Effect, I resolved to investigate, with an open mind, whether meditating could really reduce conflict in the world. Favoring a straightforward approach, I applied for a research grant from Emory University to study the Maharishi Effect by learning the TM-Sidhi Program. I was awarded the grant, and I became a Sidha (an advanced meditator) soon afterward.

At the outset of this discussion it is important for me to emphasize that I have no formal connection to the Transcendental Meditation movement. I do not represent the movement in any way. I have never received any money or other physical benefit from the movement. I report my results here entirely as an independent social scientist who has been engaged in research. All of my comments here regarding meditation reflect my own thoughts and experiences as they concern my research into extraterrestrial civilizations. I have used my own skills as a scientist and a trained observer to understand and interpret what is happening.

The Subjective Experience of the TM-Sidhi Program

In normal waking consciousness, the mind is flooded with inputs from the five physical senses: sight, hearing, smell, taste, and touch. Most of our mental activity uses information from these senses. The raw inputs from these senses feed intellectual activity such as logic and imagination. During meditation, however, the volume of these senses in the mind diminishes until they are eventually silent. At this point, logic and imagination also cease.

What remains is not an empty mind. Indeed, what remains is the mind's connection to what many meditators call a "field of consciousness," which can be very alive with activity. From a subjective point of view, this field appears both to contain and reflect the inner consciousnesses of all individuals. In the state of deep meditation, people experience this field directly. The experiential contact with this field is ordinarily overshadowed by the signals coming from the five physical senses, as well as those of thoughts, memories, and emotions. Therefore it is necessary to "transcend" conscious sensory processing by turning one's awareness away from the senses in order to perceive that which is nonphysical. Although that which is nonphysical is, by definition, not physical, it nonetheless does have an objective reality that can be accurately perceived by anyone with a sufficiently sensitized nervous system. Such perceptions are not a product of meditator dreaming.

The ultimate goal of meditation is to practice experiencing the field of consciousness until the perception of that which is nonphysical strengthens. When this process is complete, a person may no longer need to meditate. Such a person has a highly developed consciousness, and is described as self-realized. The meaning of this is that a person no longer lives a separated or dual existence in which normal awareness is limited to the perception of the outward physical senses and does not extend to the inner self. A self-realized individual has only one unified (i.e., closely interconnected) consciousness. Since the field of consciousness connects all aspects of life, one's perception of the five physical senses is then flavored by what one perceives from that broader level. Without this, one's understanding of the interconnectedness of all life is impaired. In this way, meditation infuses a sense of one's inner being into daily normal awareness. Importantly, during my investigations, I have found that the primary characteristic that distinguishes typical humans from highly evolved ETs is that the ETs are normally self-realized, whereas this is only occasionally true of humans.

As is often the case, a fully self-realized individual quickly becomes aware of the diversity of nonphysical life that co-exists with us. All of life, both physical and nonphysical, resides in a space of some sort. Unfortunately, the English term *space* (as in outer space) is restricted to that which is physical. A broader term for space, which includes both physical and nonphysical life, is the Sanskrit word *akasha*. We all exist in akasha, and all of our journeys take place in akasha regardless of whether we do them in a spaceship or by utilizing our own highly trained consciousness. Indeed, the ancient seers were the first great human astronauts, since they quite literally roamed the heavens with their minds. In general, advanced meditators and ETs similarly view our existence from this broader perspective of life in akasha.

My first experience with the TM-Sidhi Program in a very large group was while meditating in one of the two large meditation domes at Maharishi International University during the first week of 1992. There were more than two thousand meditators in the domes simultaneously with me during these initial experiences. We were all practicing the TM-Sidhi Program.

The actual TM-Sidhi Program is quite involved, and there is no need to describe any of the mechanics of the procedure here. However, it is public knowledge that part of this program involves something that is called yogic flying. There have been many public demonstrations of this phenomenon, and many of these have been televised. On a surface level, yogic flyers look like they are hopping around the room in lotus position, sort of like a frog. But the idea of yogic flying is not to do gymnastics. Rather, it is to practice doing a special type of activity while still in a meditative state intimately connected with the field of consciousness.

During the nonflying parts of these meditation experiences, I felt a distinct settling sensation (e.g., calming, sinking, awareness shifting—all of these apply in some sense). My own consciousness seemed to move somewhere, without any effort on my part. Indeed, one of the most important aspects of proper meditation is that effort must *not* occur. Effort activates the waking-state information-processing activities of the mind, which in turn overwhelm the perception of any signal from the field of consciousness. Upon arrival at this new state of awareness, there was a distinct perception of differentness between this state of awareness and my normal waking-state of awareness. It would go beyond the needs of the current discussion to describe all aspects of this differentness, but some observational comments are useful here.

In the state of meditational awareness, there was a clear sense of my own self as being different from my thoughts. Thoughts did occur in this state, but they were few in number. Moreover, I had a sense of awareness in which the thoughts were perceived as somehow alien to my essential self. Indeed, the one aspect of my perception that was most dominant was the sense of expanded awareness. From a third-person perspective, my consciousness did not think thoughts as much as it was simply aware of itself.

However, this state of expanded awareness was not without activity. I had the clear perception that I was experiencing someplace, and that my own consciousness was not the only consciousness sharing the same experience. This experience also was not momentary. Indeed, it lasted for a significant number of minutes before the program shifted to yogic flying.

From the perspective of the state of normal waking consciousness, it has often been said that the state of awareness that one experiences while meditating is very subtle. From personal experience, I can state that this description seems accurate. However, from within the state of expanded awareness associated with meditation, activities in the field of consciousness are not always subtle at all. It was during yogic flying that I first perceived what is called a "wave of consciousness" ripple through the fabric of consciousness in the domes, and it hit me like a ton of bricks.

A wave of consciousness occurs when meditators meditating in one place manipulate the field of consciousness in such a way that a surge of influence is perceived by all meditators simultaneously.

The experience is difficult to describe in words. However, in my own experience, it felt like a large flow of energy.

Of course, it is easy to dismiss such subjective observations. It is just the type of thing that one would expect an untrained participant in an experiment to say or write. But I am trained to make observations of human behavior, and I am highly sensitized to the need to distinguish between imagined and real perceptions. Personally, I cannot dismiss my perceptions of these things as they were all too real to be imagined. From my own subjective point of view, something objective was going on in the meditation domes at Maharishi International University. I do not claim to understand the physics of the experience, but it was as real to me as if I had been punched in the face (but, of course, much more pleasant).

I clearly remember my initial reaction following my first experience of meditating in a large group. I was convinced that scientists from all over the world should be in those domes with every conceivable instrument in their labs trying to figure out what just hit me. It was not physical, but it was real, and it influenced physical things (e.g., me). At the time, I became certain that nothing that strong could possibly escape the scrutiny of science forever. Yet I had no answers as to exactly what happened. In general, my mind seemed to move to a subtle level of consciousness, and from within that level I felt great activity, including powerful waves of energy.

The Monroe Institute

Twenty-one months after I completed my course in the TM-Sidhi Program, and just before beginning my training in scientific remote viewing, I attended the one-week Gateway Voyage Program at the Monroe Institute in Faber, Virginia (see Monroe 1994, 1985, and 1971). This program teaches about altered states of consciousness via brain-wave entrainment techniques. Before I took this course, I had already met the person who was to teach me remote viewing and become my monitor for the research investigations that I report here, and this person told me that training at the Monroe Institute was a prerequisite for the military remote viewer trainees who were to undergo formal instruction in remote viewing. Wanting to duplicate the known road of the military training program, I chose to attend the Gateway Voyage Program as well.

Briefly, the folks at the Monroe Institute have devised a noninvasive way of producing physical changes in the electrochemical signals that occur in the brain. Their technology is based on the use of sounds to cause various frequencies to resonate between the right and left hemispheres of the brain. Robert Monroe has labeled his patented technology "Hemisync." Hemisync sounds put one tone in one ear (for example, a 100-hertz tone) and another tone of a frequency only slightly different from the first in the other ear (say, a 104-hertz tone). The result is a very low frequency vibrato called a beat frequency (in this case, 4 hertz). Beat frequencies are not actually heard in the ears, but the mind can discern them. The brain itself *creates* them by blending the separate audio frequencies. In this way, sound is used to cause an electrochemical reaction in the brain that resonates at a frequency that is undetectable (by virtue of its low frequency) to human ears.

The Monroe Institute researchers have developed a variety of sophisticated blends of Hemisync sounds that are extremely useful in obtaining altered states of awareness by those who are listening. They have "mapped" the brain-wave electrochemical activity of many individuals who are capable of experiencing altered states of awareness. By matching the maps with the changes that occur with the Hemisync technology, they can, quite literally at the flick of a switch, make one's mind resonate mechanically like that of a great seer or mystic who has spent a lifetime exploring the boundaries of consciousness. The Monroe Institute's most interesting achievement is the discovery of a set of frequencies that allows individuals to perceive an area of nonphysical existence where life thrives.

I use a word you may have heard in *Star Trek* television programs and films to describe this place: "subspace." An aspect of each of us exists in subspace, as do other beings. People who no longer reside in our physical space "live" in subspace. It's not accurate to say that these are "dead" people, since they are very much alive.

At the Monroe Institute's Gateway course, they call the primary portal to subspace Focus 21. In this state of awareness, one can peer deeply into subspace and communicate with beings who may be there. During my second experience with Focus 21, a remarkable thing happened that greatly changed my views about extraterrestrials.

I was on my back in a small room with a bed and the necessary electronic equipment, listening to the sounds of Focus 21 through headphones. The subjective experience was one of being mentally carried to a location. The technology used requires no imaginative effort on the part of the listener. The mind automatically tunes into the frequency that the tones generate, and one is "there." After a few minutes, I seemed to arrive at someplace that had an opening, sort of like a door. The resolution of the images was not perfect, but I could sense what was going on. I was at a portal of some kind.

I went through the opening and found a room that seemed to have one wall missing. There was light everywhere, bright light. I looked around and saw one individual to my left. The being's body seemed both fluorescent and mildly transparent. The person seemed to be watching me, as if overseeing my visit. I went forward into the room and to my considerable surprise encountered three other people whom I knew. They were my grandfather, my grand-mother, and my great-aunt (my grandfather's sister).

On an emotional level, I was beside myself. Everyone was glad to see me. I was happy beyond words. I had a distant sense that there may be tears emerging from my physical body, but my awareness stayed with my relatives. My aunt (my mother's sister) had recently died, and I looked around and asked where she was. I clearly re-member asking with some concern, "But where's Elsie?"

At that moment, another—seemingly powerful—subspace be-ing covered me with what I could only think of as a dark cloak. I was then lifted up and transported to another location. During the trip, I asked why I could not see outside. (I asked this by simply thinking the question.) The moment I asked, a corner of the cloak was lifted, and I saw the brightest light I had ever experienced. It was like looking right into the sun. Moreover, I had an urge to surge into the light. The cloak was then closed again, and I under-stood. I was being protected from being exposed to the light. I had the sense that exposure to the direct light would be harmful to me while I still had a physical existence.

Shortly afterward, the journey in subspace ended and the cloak was lifted. I saw before me what seemed like hundreds of thou-sands of Greys. These were extraterrestrials, and they existed in subspace just like humans. I moved closer and extended my mind toward them. (Somehow, this seemed a natural thing to do at the time.) What I received sent me reeling.

I felt an intense sense of pain, psychological and emotional—not physical. I recoiled back into the half-open cloak. Yet I did not want to let this opportunity go, so I tried again. This time I made an effort to sense the Greys' consciousness as *they* experience it, rather than letting it make a raw impression on my own human senses. To my relief, the experience was different.

I felt a sense of calmness, quietness, and stillness. I also sensed an awareness that at the time I could only compare to the mentality of Mr. Spock on *Star Trek*. It was as if there were no surface emotions, but there was something very deep. It then became obvious to me that my first impression of pain was related to how my mind would feel if it was forced to live as a Grey consciousness. The transition process was too abrupt. I still sensed deep pain, but I was not yet decided whether it was originating from myself or the Greys. I somehow suspected both, but it was impossible to tell at the time.

At that point I heard a voice, and I knew then who had brought me to the Greys. The voice said, "These are the beings whom you want to help." It was the voice of my Aunt Elsie.

One of the Greys approached me and asked if I was here to help. I did not know what to say. I sensed that they had problems, but the initial impression was that these problems were far beyond my abilities to address.

I told the person that I did not know. I began to mumble. I mentioned that I would try, perhaps. Maybe I would be able to come back. I did not know what the task was that I should do. But whatever it was seemed overwhelming to me. I shrank, and I felt sadness, deep sadness. But I also sensed that the time to help was not now, and I must leave.

The cloak again covered me, and I was transported back to the portal and my waiting relatives. I never actually saw my aunt, but I did not really need to anymore. I said good-bye and went to the opening where the solitary subspace being I saw originally was still standing. Then, to my surprise, I saw about twenty people holding hands pass in front of me and enter a door going into what appeared to be a downward-leading tunnel. I considered this peculiar, but did not think more of it at the time.

The sound signals were beginning to change in my mind. I was leaving Focus 21. After an uneventful return, I heard the pleasant voice from the control booth.

"Welcome back! You are now fully awake and alert. Please join us in the debriefing room."

Soon afterward, the group taking the Gateway course met in the large meeting room at the institute to discuss their experiences. The first thing to come up from the discussions was that many of the participants had decided in advance to go through the experience as a group, holding their subspace hands, so to speak. They told of their voyage and their return through the door, and I knew immediately whom I had seen going into the tunnel.

As for myself, I was too numb to speak. I just listened, and wondered what could possibly happen next.

My Training in Remote Viewing

Because of the controversial nature of my research and the classified nature of the details relating to the military's remote-viewing operations, I have chosen to keep the identity of my trainer in remote viewing a secret. At some point, this person may desire to announce that he or she was my trainer and monitor for this research. But I feel it is best left up to this person to make this choice. Thus, I will simply refer to this person here as either "my trainer" or "my monitor." When language considerations force me to use a personal pronoun, I will use "he" or "him" for convenience, although this in no way implies that this person is male.

I first met the person who was to become my remote-viewing trainer (a former member of the military's remote-viewing unit) at a national conference. I talked to him at length during the conference, gave him my card, and begged for a chance to obtain training in remote viewing from him. He told me that training was not available at the time. Nor did he have much hope of beginning a new training program. Training was still organized along the original lines developed by Ingo Swann for the military trainees. In short, it was too expensive in time, money, and human resources. There were too few trainers, and none to spare.

Nonetheless, I kept this person's current phone number, even though I was told that the number was going to be changed soon because he was going to move. He told me the city he was moving to, however. Aside from some literature that this person sent me a few weeks after the conference, I did not hear from him again for over one year.

I attended that conference for only one full day and another day's lunch. My wife was pregnant at the time, and I decided not to tell her that I was going to a meeting of professionals interested in ETs and UFOs for fear of disturbing her. I do not remember exactly what I finally ended up telling her, but it was along the lines that this was a professional conference related to my research.

After I returned home from the lunch meeting, I found my wife at home, and she was quite concerned. She wanted to know what I had been doing. She told me in no uncertain terms that she had had a visitor while I was away for lunch, and she felt intuitively certain that the visitation was related to my activities, whatever they were. She told me that she had been sitting in the back yard on the bench, and that she had felt the presence of a being behind her. My wife is a Sidha and a teacher of Transcendental Meditation, so I was aware that she had refined and accurate perceptual abilities. Thus, I just stood there silently and listened to her in shock.

She then told me that this being tried to put her to sleep so that it could look at her. She felt it was interested in the fact that she was pregnant, and she did not like that at all. She resisted the urge to sleep, and instead began to turn in the direction of the being. At that point, the being moved in front of my wife, she saw it, and it then quickly dashed behind a nearby tree, seemingly disappearing. Her description of the being matched that of a typical Grey. I had never spoken to her about such things before.

My first thought was that it seemed as if some ETs were capable of monitoring conversations that some people had with others. I could only guess that someone was listening to my conversations at the conference about wanting to know more about ETs, and in particular, about wanting to write about their civilizations, and the ETs came to investigate more about me. I then told my wife everything about the meetings that I had been attending.

About fifteen months after the conference, I was meditating in my new apartment in Ann Arbor, Michigan. I usually go to the University of Michigan each summer to teach a course in nonlinear mathematics to social scientists who gather there from many countries. During my morning meditation program, I had the clear perception that I was being told to listen to one of the Monroe Institute tapes immediately after my meditation program. I had the sense that the tape should be one using Focus 12, which I have found is particularly useful for facilitating telepathic communication while remaining closely attuned with physical reality (unlike Focus 21 in which the awareness shifts almost entirely to the nonphysical realm). Immediately after I finished meditating, I went to my bedroom and quietly (in order to avoid disturbing my sleeping wife) got the tape and returned to the living room. I set up the tape in our portable player, put on the headphones, pressed the play button, and leaned back to rest and listen.

Immediately after the preliminaries on the tape were completed, I began to clearly discern the mental image of a luminous being. The person was male and was dressed in a white gown. He had a message. He told me that the time was now appropriate for me to contact the former military remote viewer whom I had met at the earlier conference and to obtain training in remote viewing. He emphasized that I should act now. The tape ended, and I wondered what to make of the experience.

Following my morning class, I went to my office and decided that I would try to contact the remote viewer whom I had met so long ago. In all of my previous phone calls to his office, I was never actually able to speak to him, as I always encountered an answering machine, and he had previously told me that he was not often there. Moreover, I did not even have his new number in the town where he said he was to move, nor did I know if he had actually moved there. With no other options, I decided to call telephone information. Somewhat to my surprise, his phone number was listed. I called, got his voice on an answering machine, left my number, and proceeded to prepare my lecture for the next day's class.

Soon I received his return call. I told him how happy I was to talk to him, and asked him if there was any chance that he might be able to train me in remote viewing as per my earlier hopes. I told him of my desire to write a book about the ET phenomenon, and I needed remote viewing as a data-gathering tool. To my very considerable surprise, he told me that he had just completed updating the training program for remote viewing, and now knew that he could train people in all of the basics during an intensive seven-day course, six hours each day. Also, it was possible for me to be his second student in this new course, beginning in about seven weeks. He told me the cost, and mentioned that he would be my trainer. The fact that he would be my trainer was important for me, since I felt comfortable with him. I told him to sign me up and arranged for my payment to him for the course. After I hung up the phone, I thought about the luminous being that had telepathically approached me during my morning session with the Monroe tape. I had the sense that surprises would continue for quite a while, and I told myself that I had better get used to it.

The timing for my training in remote viewing was prefect. In the weeks before training, I was able to attend the one-week intensive Gateway Voyage Program at the Monroe Institute. However, during that summer, I was also able to complete a home study course called the Gateway Experience that the institute also offers. It is a course using some of the same tapes used during the one-week in-residence program, and that summer I was able to spend a considerable amount of time with Focus 12 tapes. The home study course greatly enhanced my experience in the one-week residence program. The total Monroe Institute experience helped prepare me for what I was about to perceive in my remote-viewing course.

My training took place under simple surroundings. The training room was basically grey, as was the building in which it was located. There were no bright colors anywhere within the view of the trainee. The trainer wore clothing with neutral dull colors to avoid activating my visual imagination, thereby inviting contamination of the data. I was required to be well rested and not hungry during training. These are important considerations, since remote viewing exploits certain aspects of the autonomic nervous system. Anything that influences this system negatively (such as hunger) can degrade the overall quality of the data.

My trainer was an especially patient person. Apparently, it is important to build up a student's confidence in trusting the data so that it can be accurately transcribed before being changed or disregarded by a doubting conscious mind. Thus, I was particularly grateful of the care with which my instructor handled my training during those initial sensitive days.

I have found that what occurs during remote viewing is closely connected to that which happens during the practice of Transcendental Meditation, and even that which occurs while experiencing the Hemisync sound processes at the Monroe Institute. A part of the mind that is typically not utilized during normal waking consciousness is accessed and made dominant in one's mental awareness. Indeed, for a Sidha meditator, SRV is really little more than a means of accurately recording information—a transcription protocol.

The Organization of Part II

At this point, it is crucial to emphasize an aspect of the way I have chosen to represent the material in this book. In Part II, I report the results of many remote-viewing sessions made over the course of two years. The results are presented in a form similar to transcripts to ease the presentation of material. I do not present descriptions of the numerous procedures (executed both with my mind and with a pen and paper) that I did while following the protocols of SRV during sessions. *These procedures would make little sense to untrained individuals, and discussing them here would impede the sessions' readability.*

Because untrained individuals may not understand how any procedures could extract this information, some may be tempted to dismiss what is presented as coming out of thin air, and they might naturally wish that they themselves could perceive these things in order to verify the accuracy of these reports. Understand that the only reason other people cannot perceive these things is because they have not received the proper training. It is important to remember that training in SRV, while publicly available, is neither easy nor cheap. I invested a great deal of time, effort, and money to learn how to do what I report here. No one should expect these experiences that I report to be common to themselves unless they also obtain a level of training similar to my own.

The second part of this book presents the bulk of the data and interpretations that I offer. I had the choice of organizing the book thematically in terms of ET species, or chronologically with regard to my own discovery process. I chose the chronological approach, since it preserves for the reader the vicarious thrill of discovery that I felt when it happened to me.

The cost of this approach is that the chapters are somewhat jumbled from a topical point of view. In retrospect, I find this cost to be manageable, and it should bother only someone who already has an intimate knowledge of ET affairs relating to this planet. A thematic organization to a book on ETs will make more sense a few years from now, when the basic data are well understood by many people.

Before proceeding further, I suggest that readers skim the short list of SRV terms and other words in the glossary at the back of the book. These terms appear in the descriptions and analyses of the remote-viewing sessions presented in Part II of this book.

PART II

WHERE NO HUMAN MIND HAS GONE BEFORE

CHAPTER 4

My First Visit to Mars

I am in an office that my trainer uses for remote-viewing instruction. There is very little in the office to distract one's attention. The dominant color everywhere is a light grey. There is nothing on the table in front of me but my pen and a stack of paper. The weather has been perfect. The course has been going well. My last target was a bridge over a river in Viet Nam during the war. Target selection has been varied to discourage my mind from guessing at them. So far, I do not know what and where I am viewing until after the session is completed.

My trainer begins this afternoon's session like the others. He is sitting across from me at the training table. He asks me if I am comfortable, and waits until my pen is on the paper. I tell him I am ready, and he gives me the target coordinates.

Date: 29 September 1993
Place: Training office
Data: Type 4
Target coordinates: 5987/9221

I write the coordinates on the paper and then move my pen to the right of the numbers. At this point, my autonomic nervous system begins to activate my hand, and I immediately sketch a crude line drawing. This drawing is then investigated and analyzed using both intellectual analyses and my intuitions. All of this is involved in Stage 1 of the SRV protocols.

Moving to Stage 2 of the SRV protocols, I begin to collect information regarding colors, surface textures, sounds, temperatures, tastes, and smells that are associated with the target. Stage 1 and 2 procedures help me to establish a mental "lock" with the target signal. I am still new to this, of course, so it is taking me time to isolate the signal. Finally, and after eleven pages of preliminary data, I begin to bilocate.

COURTNEY BROWN: "There seems to be a mountain type of thing here. The surrounding land is a semisoft, flat, sandy area. There appears to be a sense of grandeur associated with this site. I do not see any people right now. It looks like there may be a man-made structure on the flat area."

MONITOR: "That's fine. Make your Stage Three sketch. Get it all down and move into Stage Four."

CB: "OK. Going through the matrix. . . . Things are brown and sandy here. There is a house. What is that pyramid doing here? Let me AOL [analytic overlay] on a pyramid. It must be my imagination."

MONITOR: "Don't judge things. Just put it down as an AOL for now."

CB: "The house is sort of long and narrow. Made of wood, it seems. I'm sorry, but it's that pyramid thing again. It is really huge."

MONITOR: "Keep recording your data. Put it all down in the tangibles column. What else do you see? Keep going through the matrix."

CB: "Well, there are people now. Lots of them. And animals. People and animals . . . putting it all down in the matrix. This pyramid has to do with worship of some kind. That's an intangible, isn't it?"

MONITOR: "That's right. Keep going."

CB: "The pyramid is tall, stone, hard, gritty. It is sandy and windy around here. It seems like the pyramid is solid, but hollow at the same time. Wow, it sure is tall."

MONITOR: "OK, let me give you a movement exercise. Get your pen ready. Inside the pyramid, something should be visible."

CB: "Wow. Let's see. We have browns and light tans. Surfaces are rough, sandy, gritty, stone. It is cool, but not cold. I am in a room. Let's see. There is a floor, stone walls. There is a table and a glass shape on the table."

MONITOR: "Write it all down in the correct columns."

CB: "The purpose of this place is somehow somber. There is a sense of resoluteness combined with need or necessity. Hmmm. There are tunnels. I am facing a tunnel."

MONITOR: "Go down the tunnel."

CB: "There is dirt on the floor here. The tunnel is dark. It leads outside. I am now outside of the structure on the surface. There is a road and a lot of sand around here. Again, I get the sense of somber purpose to this structure.

"Goodness, I can perceive lots of people now. I clearly get the sense that either this structure or something related nearby was a great building project, and that the folks needed help and lots of resources. Apparently, many died to build this.

"There is a nearby city. Wow. There is also a nearby mountain that is erupting. What's going on here? There are no volcanoes near a pyramid that I know of. It is like Pompeii, but there are no pyramids near Pompeii."

MONITOR: "Don't analyze. Just record the data. Keep going."

CB: "Lots of people have died, and are dying. There is a lot of movement. People are running. Many are scattered. There is a sense of hopelessness. This is terrible!"

I begin to draw a sketch of the scene. The volcano seems to be to the east of the city, and people are mostly running north.

"Moving up in time a bit, the survivors have set up a village nearby. No one was there to help them. There is desperate poverty. There are shacks and tents. This is really awful.

"Hmmm. There are some new people rebuilding the city. They are not the original folks. They are rebuilding for a new group of people, it seems. Others are coming in. These new folks come from very far away, and they do not seem to be panicky—in the sense of urgency—about helping the former residents. From the perspective of the first folks, it seems like carpetbagging."

MONITOR: "OK. That's enough for now. Write down the time and let's end the session."

CB: "Well, where was that?" *My trainer pushes a folder toward me on the table. I open it and pull out a picture taken by a satellite of the Cydonia region of Mars. There is the pyramid, clear as day. Evidence suggesting volcanic activity was to the right (east) of the pyramid.* "You're kidding. You sent me off-planet? To Mars?"

MONITOR: "Hey, I like sending my students there. It keeps them on their toes."

DISCUSSION

This was my first exposure to the idea that Mars may really have been inhabited at one time. Before this, the topic was only one that pertained to science fiction. It took me the rest of the day and evening to get used to the fact that I had witnessed an actual fragment of Martian history.

CHAPTER 5

Remote-Viewing a UFO Abduction

My trainer told me on a number of occasions that remote viewers had generally not been successful in observing human abduction experiences involving UFOs. Each time remote viewers tried it in the past, with only a few exceptions, they received a substitute signal in place of the abduction. Often the data from this signal could only be interpreted symbolically. The viewer would see something, but not the target.

In fact, the collected information might not have any resemblance to the target. When other viewers tried the same target, their results varied wildly. In some cases the abduction incident targets yielded no information at all.

The target described in this chapter is a human abduction by ETs. The targeted incident was taken from a page from Jacobs's book *Secret Life*.

My trainer knew that I would be fed a false signal in place of the abduction even though he had never given a viewer this particular target. What resulted was a signal containing symbols that the ETs wanted me to understand. Perhaps they were certain that the abduction itself would be misunderstood, so they substituted symbols that conveyed the purpose or meaning behind the abduction instead.

Of course, I was not told anything about the nature of the actual target until after the session was completed.

Date: 30 September 1993
Place: Training office
Data: Type 4
Target coordinates: 2864/0576

The preliminaries indicated that the target was on dry land.

CB: "There are earthy colors here. Browns, whites, tans. It is warm. Feels like a desert."

MONITOR: "Go to your Stage Three sketch."

CB: "Well, there is a building, a fence, and something like a railroad track."
I drew them. My trainer then gave me three movement exercises to give me different perspectives around the site.

"This feels like a place where something is kept. There is a fence here." *I drew a fence (on the paper) that curved around.*

MONITOR: "OK. Go to the Stage Four matrix."

CB: "There is a fence definitely. It is a flat dirt area. This is a workplace of some kind. There are animals inside the fence, and a few people. The people are white, and they seem to be engaged in doing their jobs. As per the animals, I detect horses.

"The gestalt here is one of controlling the animals. This is definitely a work environment, and these people are just doing their assigned tasks. They have the sense that they need to get the job done. There is quite a bit of determination with a high level of intensity here. It is like a bullfight."

MONITOR: "Put that down as an AOL of the signal line: 'like a bullfight.' Don't interpret. Just get the data."

CB: "Well, there are people in the fenced area with the animals. It feels like something is being abused. But also there are bleachers here with lots of spectators as well. People are looking into the circular area surrounded by the fence. I do not like this at all."

MONITOR: "OK, Courtney. Take a break."

Resume.

CB: "OK, I am back at the site. There is lots of shouting and screaming, also laughing. The people in the bleachers are modern-day common people. There is the sense of entertainment with regard to those people. For them, this is like a Saturday sporting event."

My trainer then gave me a movement exercise to five thousand feet above the site.

"My, this is different. There are silvery, metallic things here. There is fast movement. I am AOLing on ET ships. What do I do next? There seem to be vehicles here."

My trainer then had me execute a location-tracking technique. Using this, I followed the vehicles from above the site back to their point of origin.

"It is like the people in the ships—they look like airplanes, definitely shiny and metallic—are on a purposeful mission. OK, I have two people in a plane. They seem to have a cocky attitude. The start-off point is a city.

"There are lots of people in the city. It feels like the plane people are hurting the city people during some kind of event. The people are targets of some kind, and they do not like it. There are lots of buildings here as well. The weather is wet and cool. There is an airport where the planes take off.

"Back to the original site, the animals definitely feel threatened. It is as if the people are playing or toying with the animals. The animals appear to be panicky. But I get the clear sense that the people inside the fence do not want to harm the animals. But they do seem to be getting some entertainment from them."

MONITOR: "I want you to focus on the purpose of the people with the animals."

CB: "They are trying to control the animals, as if to herd them. This is a training camp. They are training the animals to do things."

MONITOR: "Move ahead in time a bit."

CB: "The animals are no longer panicky. Indeed, they are getting food and love from the trainers."

MONITOR: "OK. Let's end the session. Here is the target."

DISCUSSION

The best that I can offer is my personal interpretation of this remote-viewing session. Readers may wish to interpret this experience differently than do I. This type of session has only happened in two known situations. The first has been when remote-viewing UFO abductions. The second was when military remote viewers tried to observe a dangerous energy device. In both settings, someone apparently did not want humans to have access to the information, and a substitute signal was offered.

In this session, the animals seem to represent humans. The people inside the fence are ETs who are working with humans in order to train them for some purpose. The people in the bleachers are, perhaps, galactic spectators. The airplanes represent ET ships that relate to the activities of the trainers in the fence, and their occupants may support those on the ground.

This is the only target of its kind that I describe in this book. It differs from all the other targets in that *they* are actual observations and can be analyzed directly, without being understood through symbols.

Some ETs can create a substitute signal, and I now know they may tailor it to a remote viewer's mind and experiences, in a way that the viewer can best understand it. Such occurrences are very rare, however.

CHAPTER 6

Martians: Present Survivors

Still in training, I settle down into my seat. My trainer holds the closed file containing the new target. There is still the temptation for me to try to guess what the target is, and I resist this mildly. He tells me that I will soon naturally stop wanting to guess as the mind gets used to the idea that it will receive the information directly. Patience is a learned virtue, I suppose. Yesterday morning he had me remote-view this weird museum exhibit with a giant fork sticking out of a wall. I joke to myself, wondering if he is going to send me to the sewage treatment plant in Fort Meade, Maryland, that he mentions occasionally. He sits down across the table from me and asks if I am ready to begin.

"I am ready. Shoot."

Date: 1 October 1993
Place: Training office
Data: Type 4
Target coordinates: 5664/1821

The preliminaries indicated that the target was associated with a mountain.

CB: "I am getting browns and greens. It is windy and cool. I hear a swishing or whirring as well. I smell trees. Hmmm. Something is happening here."

MONITOR: "OK, Courtney, go to your Stage Three sketch."

CB: *On a blank piece of paper, I draw a rounded mountain. The top of the mountain is bald, but there are trees farther down. There is wind hitting the surface of the mountain. I move on to Stage 4.* "I am going through the matrix now. Mmm. There are people. White people. I get a sense of anticipation right now. I can see the clothing on the people. There is that mountain again, and the wind. Wow! I am picking up strong fear, also excitement and relief. It seems like there are lots of emotions here, with different people having different feelings. I am getting some kind of airborne vehicles, and lots of frenetic activity."

MONITOR: "I am going to give you a movement exercise. Get ready. From one thousand feet above the mountain, something should be perceivable."

CB: *I do the exercise.* "There is activity here, very fast activity. It is hard to figure out what is going on."

MONITOR: "OK. Go back to the people. What do you see?"

CB: "Again, there is activity, but this time it's the people who are moving. These people are excited. Hmmm. They are caught up in activity, perhaps not of their own plan. I have the mountain again. These people do have a plan they are working on, although I still cannot tell if it is their own plan. There are vehicles. I have about ten people."

MONITOR: "Let's do another movement exercise." *He instructs me to move to the top of the mountain.*

CB: "There is looping movement, very fast. It is coming down onto the mountain. It is spiraling, sort of. Like a leaf falling in the wind, or birds circling while descending in a gusty crosswind. Wow! I better AOL this. I am picking up an ET ship. Polished, metallic, and warm."

MONITOR: "Just put it down, and move onto a Stage Six sketch."

CB: "I see lots of mountains now. Many are rounded. They surround the mountain that I am on at the moment. There is a flat area to one side of this mountain, a valley separating this and other mountains in one direction. I have some plateaus off to the east, and lots of mountains around me, especially to the north and south."

MONITOR: "Another movement exercise. Inside the object, something should be visible."

CB: "OK. This is mirrorlike in here. It is shiny and polished. There are lots of lights. It is warm. I smell something that is like sickly sweet. Some type of whirring sound is here as well. Hey, this thing really is moving!"

MONITOR: "Stage Six sketch."

CB: "I have this thing going right into the mountain! Right through the rock! What is this?!"

MONITOR: "Keep in structure. Just get the data. Write it all down. Go to your Stage Six matrix."

CB: "OK, I have beings in this ship. The beings are not all of the same type. There are walls here. Devices. I am getting something from the minds of the beings. This is a supply run. No big deal. They are on a routine mission. The beings are humanlike . . . technicians. Everyone seems to have a uniform.

"I am now inside some kind of cavern or hole inside the mountain. The ship has landed in the center of the place. This seems to be a hangar or something. They do not know that I am here. There seems to be some important liquid that they are carrying. It is really disgusting in appearance, like slime. It has some biological purpose. It is an important liquid to them. It seems to have the consistency of motor oil.

"I am moving around in here now. There are lots of beings working. I can sense that the males do important work dealing with running these operations. The females do not work in the technical areas. They seem to do other work. It is like they are cared for, and that they do less-important work.

"Still moving. There are children here. The children are not well. These kids are sick, real sick. The females are almost in a state of panic—they are barely containing themselves. They are sitting quietly, but wow are they upset. Very frightened. I am picking up the males again. You know, this culture seems sexist."

MONITOR: "Let's take a break, Courtney. Write down the time."

Resuming after lunch.

CB: "Well, I am back with the women in the nursery. There are containers for the babies. The kids are not talking. They are either somber, or sacked out, or unhappy. Something does not feel right here. I see some male and female adolescents. The adolescents seem to be OK. But there are not many of them. Lots more babies, and they are sick, or at least most of them. The young people are ignorant of the problem. But the moms seem to know what is going on.

"The physical environment is the thing that is not healthy. It is the total ambience that's the problem, not only the technical malfunction of the body. It seems like they need to break out of a culture or social bond, like they are in jail. The situation needs some new element or ingredient. I am getting the sense that something human would help here."

My trainer then has me focus on the solution to the problem.

"There is a genetic problem. It seems like genetic changes made to their own bodies are still being done. I am getting it clearly now. These beings give me the feeling of Martians, live ones. They can't fix the genetics. This is a tremendous problem for them. There is widespread desperation.

"Their equipment and resources are not advanced enough to solve the genetics problem without outside help. As far as the women are concerned, there seems no way out. They just sit here hoping for the men to get it together. The men have a very narrow focus on their activities. They are angry and stubborn. Survival is the key here. Survival. Gosh, these beings are desperate!"

MONITOR: "Courtney, find out more about the liquid."

CB: "The stuff comes from Mars. Maybe I should put that down as an AOL. . . . Who knows who these people really are. I just have the sense of Mars with these people."

MONITOR: "Keep to structure. Just put it down. Don't analyze."

CB: "Well, the liquid is ugly. It tastes bad and is gross. But these folks value that liquid like it's their lifeblood. It is in large canisters.

There are environmental units that preserve and protect it. The stuff is greenish black."

MONITOR: "Go to where the liquid is produced."

CB: "Wow! Where am I now? I just got catapulted somewhere. It felt like a whipping action, like I snapped to some other place.

"This place is red, sandy. There is a structure. I can go inside the building. There seems to be some sort of sealed door. Should I go through the door?"

MONITOR: "First tell me more about the environment around the building, and then go into the building."

CB: "It's a desert. There is nothing growing out here. Barren. Cold, too. The building is like an adobe house. Inside there is metallic and plastic surfaces. It is shiny. It is a production facility."

MONITOR: "Let me give you a movement exercise." *Pause.* "Five kilometers east of the production building, something should be visible."

CB: "Hey, we have one of those ships again. It is doing that crazy movement, curving and looping, as it descends. It went right into the building from the top. It went right through the roof!"

MONITOR: "Go back to the building and follow the vehicle. Where does it go?"

CB: "I'm in the building now. Hmmm. The ship went down. There are many underground chambers below this building. The beings in the ship do not like to walk outside of the structure. There sure are a lot of reds and tans outside of this building. And I keep getting the sense of Mars."

MONITOR: "Go into the underground chambers."

CB: "The place is modern, but not supermodern. I see men here, no women. There are workers here. This is not a happy work environment. These people are here out of duty. I am going further down.

"They live down here. Virtually a city. There are many caverns and tunnels. Machines are all over the place. It is more comfortable here than in the work caverns above, and these people could live here a long time. I detect a fear of leaving this place."

MONITOR: "Why are they leaving?"

CB: "There is no future for them here. This is a dead place."

MONITOR: "Describe what the people look like."

CB: "Well, I see males now. They have humanoid faces, but no hair. They do not look exactly like normal humans. It is like they are a different race. They seem to have mental machines as well, like devices that interact with their consciousness in some way. Their minds control the devices. The beings themselves are light-skinned. They also seem rather weak relative to humans."

MONITOR: "OK. Let's end the session. This is enough for now."

CB: "Phew! That was a long one. All right, tell me. Where was I?"

MONITOR: "The sewage treatment plant in Fort Meade, Maryland."

CB: "Huh?!"

MONITOR: "Only kidding. Here is the folder. Take a look." *I open the folder and pull out a piece of paper with these words written on it: "Martians/present survivors." There is a long pause.*

MONITOR: "Are you OK?"

DISCUSSION

Later that day, my trainer and I talked at great length about Martian matters. He told me that he had an idea about the location of the mountain based on the descriptions that I and other remote viewers had given of it. He suggested that I look at some pictures of some mountains near Santa Fe, New Mexico. When I did this, I had a sense of inner recognition. Subsequent remote viewing by a number of other viewers tended to confirm the location. The remote-viewing evidence suggests that the mountain is Santa Fe Baldy, located inside a national forest not far from Santa Fe, New Mexico.

There *are* Martians on Earth, but one must think clearly about the implications of this before ringing the alarm bell. These Martians are desperate. Apparently they have very crude living quarters on Mars. They cannot live on the surface. Their children have no future on their homeworld. Their home is destroyed; it is a planet of dust. In the final chapter of this book, I review my ideas regarding how humans should respond to our Martian neighbors in this time of great need.

CHAPTER 7

Martian Civilization: Apex

Training has been going great. Yesterday afternoon, my trainer had me remote-view Monterey Bay, California, where I ended up peering down on a sailboat. This morning, however, the session is of a different type. He wants me to experience a session under Type 6 conditions. This is where both the monitor and the remote viewer are front-loaded with significant target information.

Since I was to know about the target in advance, I've chosen the time and place that I wanted to see: the Martian civilization at its peak. I wanted to see what kind of society they had before the collapse. (In a later chapter I describe the catastrophe that caused this collapse.)

No amount of guessing could have prepared me for what happened in this session. One of the many things that I was to learn from this session was just how important an experienced monitor could be when unexpected things occur (which is to be expected)!

Date: 2 September 1993
Place: Training office
Data: Type 6
Target coordinates: 8587/7258

The preliminaries indicated that the target was associated with dry land and artificial structures.

CB: "I am getting browns, tans, and reds. It is sandy and windy around here. The temperature varies from warm to cool. I hear voices, music, talking. I also hear some type of rubbing and hub-bub. The ambience of the place feels a bit like Old Town Mombasa, the ancient Swahili port city on the east coast of Kenya."

MONITOR: "Go on to your Stage Three sketch."

CB: "I have a road here, with buildings on one side. One person is standing near a round structure. It gives me the sense of a small amphitheater."

MONITOR: "OK. Put that down as an AOL for now: 'like an amphitheater.' Move on to Stage Four."

CB: "I am now picking up people, lots of people. I now see only men. Focusing on their faces. They have no hair and have larger eyes than humans. Their skin is light. There are houses. The buildings seem to be made out of clay or adobe. These folks are poor by current Earth standards, but they seem happy. It appears to be a harsh place to live in general.

"There is lots of water around here. These folks like water, it seems. They have basic tools. Their means of communication seem to be rather basic as well. This reminds me of Africa."

MONITOR: "Put that down as an AOL: 'like Africa.' "

CB: "Focusing in on their minds. They do have some telepathic abilities. OK, I have located the women and children. The women are inside the houses for the most part. They don't go outside much with the children."

MONITOR: "Can you detect anything about their culture?"

CB: "Well, they seem to have meetings, sort of like village meetings. Let me look at the men again."

MONITOR: "Let's take a break."

Resume one-half hour later.

CB: "Well, I have the houses again. Going into one. There are three rooms. There is a toilet in the house. Folks around here feel this is a comfortable life. I see utensils, cups. A family lives here. OK, I have four people, male and female. I get the sense that these people are polygamous."

MONITOR: "See if you can discern any type of symbol."

CB: "Uh-oh. What just happened? I have experienced a time dislocation. It felt like I was snapped to another period. It was sort of a whiplash effect. Things are different here. What is going on?"

MONITOR: "Don't analyze. Just get the data. What do you have?"

CB: "I am staring at an insignia. There are polished white surfaces around here. I see metal around here, also grey and black smoke in the air. There has been an extremely rapid technological acceleration compared to where I was last.

"I see other beings here now. They are smaller, shorter. They give the sense of being workers on a mission. Wow, are they motivated. For some reason, speed and urgency are paramount in their minds.

"These other beings have ships, spaceships. They have uniforms with insignias. Some of these beings are pilots. I do not see any Martians right now."

MONITOR: "Try to find out where the Martians are."

CB: "That's just it. The Martians are gone. They're finished. The houses are empty. I am still on Mars, but it is a ghost town except for these short, advanced beings.

"The short beings have set up their own houses. They are modern, boxlike. There are technological devices in the houses. I get rooms, again modern."

MONITOR: "Focus on the purpose of the shorter beings."

CB: "They are here as the first stage in a larger project. I get the sense that they are packaging everything."

MONITOR: "OK. Let's do a Stage Six timeline. Place the apex [peak] era on the line." *Pause.* "Now locate the point on the line where you have the arrival of the others."

CB: *I place the apex period on the left side of the timeline and the time of the arrival of the others halfway across the page to the right. My trainer has me spend a considerable amount of time drawing the insignia that is on the uniforms of the short beings. It is shaped like a Valentine's Day heart with a coiled snake in the center. The border of the heart shape is gold, the inside background is white, and the head of the snake is red.*

MONITOR: "Good. Now go through your Stage Six matrix."

CB: "I get the sense of two different types of beings. The Martians themselves thought of these folks as being of a different tribe of Martians, not necessarily space people. The Martians just did not understand.

"The situation was one of panic and despair when the short beings arrived. These short beings are milky white. The Martians viewed them as godlike. I am picking up a red liquid. The short beings are using this liquid in some way. Somehow the Martians are getting packed up in preparation for some change. This is weird. It feels like the short beings are planning on the Martians getting a physical change in their bodies, and they're being put in cold storage for a while.

"These little short folks look like Greys."

MONITOR: "OK, Courtney. Let's end the session. Put down the ending time."

DISCUSSION

My trainer's cue to search for a symbol sent me in a completely unexpected direction—through time—to an insignia on a Grey's uniform. I have since had many such experiences. The sensation is similar to a rapid physical movement, but it is not possible to confuse the two. You feel sudden acceleration followed by stillness, and a momentary sense of disorientation.

At its peak, Martian society compared technologically to that of ancient Egypt. They were a people who lived under harsh conditions. But they could feed their families, live in towns, participate in community life. Men and women performed very different functions in that society. It was not an egalitarian society. Women stayed in their houses with the children for the most part.

Interestingly, this cultural aspect does not seem to have changed much today.

Martian society experienced some major catastrophe. Many Martians died and some were rescued, although I am not sure the Martians liked the terms of the rescue.

The rescuers were the beings that we now know as Greys. They arrived at the last moments of the collapse of the Martian civilization. With great speed, they somehow "stored" the Martians. Apparently this was necessary to preserve what could be saved of Martian life. I cannot provide a technical explanation, but the essential focus of the rescue was the preservation of Martian genetic material.

All of this happened millions of years ago. After this session, my trainer and I wondered how the Martians became "unpacked," and how they arrived on Earth. Virtually all remote-viewing data show that the Martians seem to have been genetically altered to enable them to live in the heavier gravity and different conditions on Earth. The actual alteration occurred recently, following a period of preservation, and is not yet complete.

We also wondered about the Martians' technological advances. Today's Martians have advanced technology, but not in comparison with the Greys. We now know that the Greys have technology that can enable their spaceships to traverse both time and very long distances, that is, distances on a galactic scale. The Martians cannot do this with their spaceships—otherwise they would return to their own planet at a time before the catastrophe. Their ships do use advanced (by human standards) propulsion technology, and they can alter the matter phase of their ships, enabling movement through solid material.

From this session, we drew some very preliminary conclusions: (1) The Martians were rescued from total extinction by the Greys. (2) The Martians were delivered to the modern time period with genetic alterations that were not perfect, apparently resulting in the deaths of many of their children. (3) The Martians have been set up with a level of technology that appears to be about 150 years ahead of current human technology. (4) Modern Martians have no place to seek refuge other than Earth.

At this point in my research, I began to wonder if there may be a reason behind the Martians' modest advantage in technology.

Someone seemed to have set things up with the potential for a major transformational interaction between humans and Martians, with each needing the other.

Recall that the Greys came to the rescue of the Martians only at the last minute. If one were to predict the future based on past data, one might foresee a catastrophic crisis here on Earth. Such a crisis would force humans to reach out for help from wherever it can be found. Martian technology might be exactly what is needed in such a situation.

CHAPTER 8

Subspace Helpers

Same day, 1:30 P.M. My trainer and I just got back from lunch. We went to this incredibly good and inexpensive healthfood-vegetarian–no-MSG Chinese restaurant. The exotic lunch was actually useful in terms of breaking my mind away from the intensity of the morning's session with the Martians. I was beginning to realize that the ET situation was much more complex than I had previously thought. It was no longer a simple situation of ETs flying around Earth. There was real purpose behind their activities, and I knew that at least some of the Martians were in considerable difficulty, and may have been for quite some time.

We wondered how to help the Martians. They had long lived underground, hidden from the harsh conditions of their homeworld's environment and from human hostility on Earth. The Martians did not have the resources to better their situation, but neither of us could tell exactly what they needed—just that they needed help soon.

This chapter presents an SRV session in which my trainer chose the target and I was blind.

Date: 2 September 1993
Place: Training office

Data: Type 4
Target coordinates: 8976/6643

The preliminaries indicated that the target was associated with complex man-made structures.

CB: "OK. I am getting lots of colors. Blues, reds, primary colors mostly, blacks, greens. The textures are paintlike. Smooth, polished, shiny. I hear air passing. It is warm here, comfortable. Hmmm. This place feels advanced. It feels like I am new to this place, and I am feeling a bit awkward, like I both shouldn't and should be here at the same time."

MONITOR: "Go to your Stage Three sketch."

CB: "All right, let's see. I have something black and rectangular. Movement on top. Lots of rectangular objects, large. Heck, this feels like a city."

MONITOR: "Put 'city' down as an AOL for now. Move on to Stage Four."

CB: "I have multiple structures here. Buildings are all over the place. I am picking up a sense of purpose here, like there is a goal that needs to be reached. I am also getting a sense of excitement from something or somebody. And . . . this is weird."

MONITOR: "Don't analyze. Just go through the matrix. Write down your data."

CB: "But it is just that I am picking up that this target has a purpose for me. I never felt that before. I am getting the sense of somebody here, nonphysical."

MONITOR: "OK, let's take a short break for now. Put down the time."

Five minutes later

CB: "Well, I am back at the buildings now. There are other beings here. I am detecting a sense of purpose among them."

MONITOR: "Where do you sense you should be?"

CB: "I feel like I am supposed to go after the buildings first."

MONITOR: "OK, start with the buildings. Continue through the matrix."

CB: "I am at the buildings. I sense that this place has something to do with a purpose of rescue, like it is a workplace where that type of work goes on. I think I am supposed to go into the buildings."

MONITOR: "Move inside then. Keep going through the matrix. Write down everything."

CB: "Wow! There really are beings here. These are *not* human beings. You can see right through them. What is this place?"

MONITOR: "Stay in the matrix. Do not analyze. Move quickly through the columns of data. Keep going."

CB: "Well, I am in this room. There are walls, and there is light coming from the walls. There is white light. I now sense that I may have been here before, but I do not know when.

"Uh-oh. I am being given a welcome. These beings know that I am here. They are looking right at me. I am getting very nervous about this."

MONITOR: "Stay in structure. Continue through the matrix."

CB: "There are doorways. These beings live here. They do work here. There is a table in the room."

MONITOR: "What about outside of the building? What do you see?"

CB: "This is a city. There are streets, lots of streets. It is noisy outside."

MONITOR: "Go back into the room. What work are the beings engaged in?"

CB: "These beings are radiating a sense of excitement, perhaps due to my arrival. It is like they were expecting me to come. One of them eagerly wants to tell me the answer to the question. It seems that they work with people. My gosh, I am being told that they work with souls. These are very advanced beings. There is a lot of excitement around here. I guess having remote viewers drop in on them like this is not an everyday event."

MONITOR: "Keep in structure. What is their work?"

CB: "There is a lot of light around here. They do not work with physical tools. I am being told that their purpose is to go forward, to move on, in the sense of evolution."

MONITOR: "OK. Let's take a break for a while. Stand up and stretch. We can walk outside a bit."

Twenty minutes later

MONITOR: "Find out more about the projects of these beings."

CB: "Well, it seems that they have been called angels in the past, but they are not angels. I am being directed toward a hallway and other rooms. There are other beings here. But there are also human etheric or subspace bodies here as well."

MONITOR: "Find out what they are doing with regard to the future."

CB: "It seems like they are here out of necessity. Hmmm. I am being told that different races or species are going to be involved. A bad time is coming. There will be a period of great struggle. During this time, technology will grow slowly. People will go back to basics in living, but not primitive."

MONITOR: "Ask about the Martians."

CB: "I am being told that humans will meet with the Martians near their [the Martians'] home—near the New Mexico caverns. The Martians have great fear. We [humans] must help bring them out of the caverns.

"These beings are telling me that we have to be both assertive and passive in getting the Martians out. We need to be very clever. This is not a simple task. The Martians do not want to come out. They fear the aggression of humans. Apparently, we are not so civilized from the Martian perspective. Yet we need to talk to them, negotiate.

"OK, I am being told something very directly. The Martians will need formal talks."

MONITOR: "Where will the talks be?"

CB: "In a house, a human house. Humans will be able to enter the caverns only when the Martians are ready to leave, not before. I am being told that we need to make assertive contact ourselves. We have to go after them. We must not stop trying. However, we need to proceed one step at a time. I am being explicitly told that we will not be hurt by the Martians. We are to go to them, not to expect them to come to us."

MONITOR: "How are we to proceed?"

CB: "We are to begin by training more people. It seems like training is a big thing. Remote viewing is a part of it, but there is more."

MONITOR: "That is enough for now. Say 'thank you' on your way out. Let's end the session. You can take a look at the target now." *He slides a large manila envelope over to my side of the table. I open it and pull out a piece of paper. "The Midwayers," it reads.*

CB: "Who the hell are the Midwayers?"

MONITOR: "It's a long story. Why don't we begin with some background? . . ."

DISCUSSION

Over dinner, my trainer filled me in on the background of his interaction with the Midwayers. In the early years of the military's remote-viewing investigations, some members of the remote-viewing team wanted to investigate certain nonphysical targets, one of which originated from *The Urantia Book,* a book of spiritually oriented revelations. The team targeted a group of subspace beings called the "Midwayers." According to *The Urantia Book,* these beings never assume physical form, although their density is close to human physical density and their bodies are just out of range of our physical abilities to perceive them. These Midwayers are assigned to Earth to assist humans in matters dealing with human spiritual evolution.

The discovery that the Midwayers actually exist was a shock that reverberated through the consciousnesses of the military's SRV team for quite a few years. On one hand, the information of their existence was incredibly important—for all sorts of reasons. But on the other hand, it was never clear how one could explain all of this to the generals who were more concerned about warhead counts in missile silos.

The Midwayers are themselves not extraterrestrials, since they apparently are rather permanently based here on Earth. Yet they are not human, nor do they assume human form in a physical sense. They are subspace beings who live and work in a human environment.

The Midwayers work here, but their command structure does not originate on this planet. Apparently, they are one of a number of subspace groups that are assigned a variety of tasks. The Midwayers work together as a unit, in the way that a military team works together. But they are not militaristic. They work with the subspace aspects of humans to promote human evolutionary potential. In a very real sense, they seem to be "good deed doers," and I do not yet understand in any depth their motivation for helping us. They seem to be working for some goal that is important to themselves and others, including humans.

My trainer introduced me to them in part to give me experience working with telepathically capable subspace beings. He did not know where they would take me, nor did he know what type of information they would give me.

It was quite some time before we could figure out just how we could help encourage the Martians to come out of their caverns. The big trick seemed to be to get them to begin formal talks with humans. The Midwayers left the clear impression that we should be assertive with the Martians, without threatening, and this seemed like a potential contradiction. A number of weeks after my initial training in SRV, it came to me that one way to help initiate a desire among the Martians to begin working directly with humans would be for there to be widespread knowledge of their activities, including the geographical location of their homes in their caverns, in which they live to avoid being discovered by humans. If humans were to find out about them, I reasoned, and if humans were capable of repeatedly locating them and following their movements and activities, they would no longer have reason to hide. Indeed, under such circumstances, the only reasonable choice of action would be to open up negotiations with humans.

But there are two sides to this situation. It is one thing to get the Martians to want to deal with humans. It is quite another thing to get the humans to want to interact with Martians, and both my trainer and I feared that the latter problem was going to be the more difficult one to solve. We sensed that we needed help with the human side of the equation. Yet we both felt deeply on an intuitive level that neither one of us would have gotten this far if there was not some hope of a solution. In a sense, we felt that there was someone else watching what we did, and that somehow the resources would be made available to us when the time was right. For now, we could only proceed, and to move forward at this time simply meant to collect more data and to proceed with my evolving plan to write a book on the subject.

CHAPTER 9

Shot from the Sky

On 21 August 1993, a NASA space probe was approaching Mars when suddenly all contact was lost between the probe and ground controllers (*New York Times*, 24 August 1993, p. A1). The probe, *Mars Observer*, was scheduled to take detailed photographs of much or nearly all of the Martian surface, including areas in which previous satellite photographs showed surface anomalies, which arguably represented pyramid structures and facelike geological carvings. Following the unexpected silence from a previously flawlessly working satellite, technicians and scientists at NASA were at a loss to explain the situation.

In the days immediately following the event, the *New York Times* reported that some folks at NASA were openly wondering why Mars seemed to be jinxed. Among other mysterious occurrences relating to Mars, a Soviet probe died under similar circumstances as it approached one of the moons of Mars not long before this. Some people in the agency actually wondered aloud—only half-jokingly—if there could be an extraterrestrial cause for the series of curious technological failures involving that planet. After months of investigation, the agency announced that the probe probably blew up because of some internal fueling problem. But the investigators were not sure about this diagnosis, and there were no data to support this claim. It was a hunch, but it was the best they could do at the time.

This chapter explains what actually happened to the *Mars Observer.* Readers should be reminded that no prior information was given to me regarding the nature of the target either before or during the session. Moreover, the data were collected in a remotely monitored session. That is, the session was monitored from my monitor's home while I sat in my office at Emory University in Atlanta, Georgia. Such monitoring is both verbal and visual. Speakerphones are used on both sides to keep the monitor and viewer in constant contact. Moreover, intermediate results (including sketches and raw data) are faxed to the monitor during the session, as are final results following the completion of the session. As with all Type 4 data, the viewer is given target information only after the session is completed.

Date: 7 February 1994
Place: Atlanta, Georgia
Data: Type 4, remotely monitored
Target coordinates: 6421/9054

My Stage 1 data indicated that I was dealing with a hard, man-made structure together with the sense of movement.

CB: "I am getting a lot of movement here. Something is going very fast, highly energetic. Hmmm. I am getting two objects, together, or at least very close. One is small, hard, solid. It is moving very fast. The other is a larger, more complicated, irregularly shaped object.

"This is weird. There doesn't seem to be any ground underneath either object. I do not know why I cannot see the ground. They are just there."

MONITOR: "Move on to Stage Six, matrix and sketch. For your sketch, place an X on the paper to represent the location of the objects. Trace the movements of the objects."

CB: "Well, the smaller object came from off to the side. I am following it back now to its point of origin. Uh-oh."

MONITOR: "What do you have? Stay in structure. Go to the matrix."

CB: "It's a ship. The small thing leads back to an ET-type ship. It was apparently shot like a projectile out of the ship and it hit the other, larger object, the one that was irregularly shaped. Why would they do something like that?"

MONITOR: "Don't analyze. Just collect the data. What do you see?"

CB: "Well, I am going inside the ship now. Hmmm. There are beings in here. They are bald, all of them, it seems. They have eyes. I am sketching the face of one now.

"The entire ship seems like a large metal structure. I am in a room. There are things in this room, lots of things, technical things. Chairs, tables, a few beings, computer terminals, things like that."

MONITOR: "OK, let's take a break for now. Write down the time, fax me the results so far, and call me back. See you."

CB: "Right. Give me a few minutes."

Resuming

MONITOR: "Courtney, go back to your Stage Six sketch. I want you to trace the movement of the ship back to its origin."

CB: "OK. Doing that now. . . . OK. I have the starting point. It is a hole in the ground, a cavern. The metal vehicle is in a cavern. Beings are getting off and on the vehicle."

MONITOR: "Go to the surface. What do you see?"

CB: "I am going up now. There are reds, sandy textures, rough terrain. This seems like Mars."

MONITOR: "That's analysis. Put that down as an AOL for now. Go through the matrix. Data only. Go back into the cavern."

CB: "Well, there are beings in this cavern, lots of them. They are Grey types. They are working."

MONITOR: "Courtney, I want you to pick one of the beings and go into its mind. What do you get?"

CB: "OK. I have one. Wow!"

MONITOR: "Put that down as an aesthetic impact—AI. Go on. Find out something about them. Find out if they sleep."

CB: "Gosh. I am getting this clearly now. The Grey knows that I am here. It does not seem to mind my probes. Very natural feeling. It may not sleep like we sleep. Something else happens. Its comparable process is when its consciousness goes very deeply back. I am not exactly sure what this means. Should I follow the consciousness back?"

MONITOR: "Go ahead. Keep in the matrix."

CB: "Wooo. It is deeply, deeply black. It is a void—emptiness, space. It is not bad, but I do not know what to do here. What should I do next?"

MONITOR: "Follow the being through time to its point of birth. Where did it come from?"

CB: "I have it now. The infant is in a tube, like a clear canister. I am at the new location now. I do not know where this is, but it looks like a laboratory."

MONITOR: "Go outside. What do you see?"

CB: "This is an airless world. I see stars, craters, rocks. I'd better AOL this impression—it is like the Moon. The light is incredibly bright here. Lots of stars—so bright! Boy, are things clear here. Looking around. There is a planet in the sky. Hey, I am sorry, but that thing looks like Earth. I can even see the clouds and water. The thing is blue. OK, let me AOL that as Earth."

MONITOR: "Go back to the infant in the tube. What is in the tube?"

CB: "Just the infant, like a big fetus, and a thick liquid. The liquid is green."

MONITOR: "Taste the liquid. What does it taste like?"

CB: "Yuck. It tastes terrible, like oil."

MONITOR: "OK. Return forward in time to the being in the cavern where the ship was. Find out more about the working environment and the being's personality."

CB: "This Grey—let me call it a Grey since that is what it looks like—is not happy by our standards. It is working.

"I'm moving into its mind now. It seems emotionless. I even get the sense it has been psychologically raped. This is not good.

"It seems to know little else. I do not have a good feeling about this being. Something is not right here."

MONITOR: "Sketch the being."

CB: "Got it. The skin is white and leathery. The being actually seems quite strong despite its thin appearance. But somehow I feel sorry for the being. It is not a good situation for it. I am just feeling bad."

MONITOR: "OK. We have to end the session now. You are starting to empathize with the being, which can begin to taint the data. But this has been very good up to now. Put down the ending time."

CB: "Well, this session sure is a mystery to me. I can't imagine what the target could have been. What was it?"

MONITOR: "It was '*Mars Observer*/loss event 1993.' "

CB: "You're not joking?"

MONITOR: "Nope. It was the *Observer*."

CB: "So that was the irregularly shaped object! It was hit by a big bullet? Why the heck would they do that?"

MONITOR: "That's a good question, but we have to believe the data; this session was pristine. Apparently they did not want that satellite around taking detailed photos of . . . whatever. It may seem strange to use a cannon-type device to take it out, given this day and age of lasers, et cetera. But remember that the Soviet probe also died of mysterious causes, and the last telemetry information from that probe was a visual image of an approaching object or energy source combined with a shipboard energy surge. My guess is that the ETs did not want any chance of a similar leakage of data, so they took it out physically. For all humans would ever know, it could have been hit by a meteor."

CB: "I am still a bit numb. I can hardly believe it, but it all fits. I am just repeating the target in my mind, waiting for it all to sink in."

MONITOR: "Good session."

CB: "Yeah. Let me get off now so I can fax you this. We will talk later tonight. I just can't get over it."

MONITOR: "Good plan. I am waiting for your fax. Talk to you tonight. Take care."

DISCUSSION

A great deal of information came out of this session, and it is useful to summarize the primary impressions. The *Mars Observer* was destroyed by a projectile-type device that was launched by a nearby ET ship. The ship originated (or returned) to an underground hangar, apparently on Mars itself. There were many beings in the hangar, all in a state of great activity. Some of the beings (but not all) were small Grey types. They were workers. I "followed" one of the beings back to its birth, and found it was "born" in a laboratory. It may have been created to be a worker. The being itself does not necessarily feel that it is being abused or enslaved. During periods of apparent sleep (in a Grey sense of the word), the being's consciousness resides in someplace that seems empty, spacelike—a void. It does not seem to dream. The laboratory in which it was born seems to be on our Moon in an underground structure, a base of some sort. The nutrients that surround the fetuslike infant are in a green liquid that has the consistency of motor oil.

This session raises as many questions as it offers answers. We now know what happened to NASA's Mars probe. But we still do not know what it was that the ETs did not want the satellite to detect or photograph. It is not clear whether the Greys involved with the base on Mars were working with Greys elsewhere. In this situation, the Greys seemed to be workers, and other humanoid beings seemed to be in control. Unfortunately, I did not clearly perceive exactly who was in control.

CHAPTER 10

The Galactic Federation

The extant UFO literature based on abductee reports often refers to an extraterrestrial organization called the "Galactic Federation." It is supposed to be an organization somewhat similar to the United Nations, except on a galactic scale. The current remote-viewing session was designed to find out more about such an organization, should it exist. The results of this session were so surprising to my monitor and me that I present it here with very few introductory comments. This is Type 4 data, and readers should remind themselves that I did not know that I was remote-viewing the Federation until after the session was completed.

Date: 9 February 1994
Place: Atlanta, Georgia
Data: Type 4, remotely monitored
Target coordinates: 3114/0029

The preliminaries suggested that the target was associated with an artificial structure, movement, and a high level of energy.

CB: "I am getting a lot of energetics here. This signal seems particularly strong. There are very bright lights—yellows, whites, blues.

There is a lot that is airy here, smooth, even fluffy. I am getting both hot and cool temperatures, and a sense of expansive and radiant energy."

MONITOR: "OK. Move on to your Stage Three sketch."

CB: "I am drawing something hard and round in the center, surrounded by bright light—yellow, blue, and white light. There seems to be something fluffy around the light, almost like clouds. The thing in the center may either be metallic, or have something metallic in or on it. This entire thing gives me the AOL of a tornado, since I am getting the sense of a swirling around the hard center thing, like a vortex. It is like this is an energy vortex. Tremendous energy."

MONITOR: "OK. AOL the tornado, and move on to Stage Four."

CB: "Again, I am getting a lot of light. Something is round and circular. I am also detecting a consciousness. I am getting some type of spiritual sense."

MONITOR: "Get that all down in the matrix, then let's take a break. Fax me your data up to this point and then call back."

CB: "Right. Talk to you in a few minutes."

Resuming

MONITOR: "Courtney, move on to the surface of the hard object and then continue with your Stage Four matrix."

CB: "Doing that now. . . . Wow! I have to AI this. Very powerful energetics."

MONITOR: "OK. Let the impression fade, and then move on."

CB: "Again, there is that blue and airy white light. Again, lots of energy. I am on the surface now. Hmmm. It is a place. The round object may have been a planet. It is misty, foggy here. The light show is mostly above. It is sort of cold on the surface. I am getting a bitter taste of ammonia in the air.

"I am seeing things that go up. Let me move closer. . . . I am now next to something that is hard and metallic. It is a building or structure of some type. This structure has some purpose for some type of beings. I see an opening, perhaps a door. Should I go in?"

MONITOR: "Before doing that, move back a bit so you can see the structure more completely."

CB: "Doing that now. . . . It is really pretty big. *Towering* is a good description. In fact, as I move around, it is humongous. It seems made of metal. I see no other nearby structures, in the sense of a city."

MONITOR: "OK. Now go into the structure."

CB: "I am back at the opening now. Going inside. . . . OK, there are beings in here, lots of them. This place is weird. These beings are all bald."

CB: "Keep in structure. Be careful for AOLs. Fill in the matrix."

CB: "These beings are all wearing white gowns, like nightgowns. Their skin is very smooth, either white or off-white complexion. This feels like an important place."

MONITOR: "Sketch the face of one of the beings."

CB: "Doing that now. . . . They are humanoid. This place reminds me of a Zen monastery."

MONITOR: "Put that down in the correct column: 'like a Zen monastery.' Keep going."

CB: "These beings use telepathic and verbal communication. I get the clear sense that this is some type of council. There is a centralized organizational structure to this council. They do not seem to be aware that I am watching. They seem to be concerned with matters of state or governance.

"I am focusing a bit more on the members themselves now. These folks *wanted* to do this job. This is a highly desirable job, and it is very competitive to be here.

"I am now picking up that there is a head council member who directs activity. The others support him. He is sort of like a president, chairman, or prime minister. Uh-oh."

MONITOR: "What's happening?"

CB: "Looks like I was wrong about them not knowing I am here. I am being given a welcome. I am being told that they are glad that *we* have arrived. *We* are now on the council. They are telling me that human representation on this council begins now.

"I am now being brought to the head being. He's looking me straight in the face. He is sitting on a chair, and he has a white, or bluish-white, gown. He seems a bit heavyset."

MONITOR: "You are on your own now, Courtney. Just keep in structure. Write everything down."

CB: "The fellow is somber, but a definite sense of humor is leaking through. He is totally nonthreatening to me. You know, this feels like going to see a spiritual master, like Buddha."

MONITOR: "Put that down in the AOL column. Follow his cues."

CB: "He is welcoming me right into his mind. He actually wants me to enter his mind as a more effective means of communication. What should I do?"

MONITOR: "Go in. Cue on the word 'guidance' and see what he does."

CB: "As soon as I went into his mind, I re-emerged in space. That is where I am now. I am outside of the Milky Way, looking onto it. Dotted lines have been drawn over the image, dividing up the galaxy—like quadrants.

"I am being told that there is a need for help. They need us. I am getting the sense that they need us in a galactic sense, but I seem to be resisting this. They are so much more powerful than humans; it just does not make sense why they would need us.

"The leader is sensing my resistance and redirecting me to a planet. OK, I can see it is Earth. I am being told that there will be a movement off the planet in the future for humans. I am just translating the gestalts now into words. But the sense clearly is that Earth humans are violent and troublesome currently. They need shaping before a later merger. Definitely humans need to undergo some sort of change before extending far off the planet.

MONITOR: "Ask if there are any practical suggestions as to how we can help."

CB: "I am being told in no uncertain terms that I am to complete this book project. Others will play their parts. There are many involved. Many species, representatives, groups."

MONITOR: "Ask who else we should meet using remote viewing, or another technique."

CB: "Only the Martians. Hmmm. I am being told that our near-term contact with extraterrestrials will be limited to the Martians for now, at least in the near future."

MONITOR: "Ask if there is new information that we need to know but do not know now."

CB: "This fellow is very patient. He knows this is hard for me. He is telling me that many problems are coming. There definitely will be a planetary disaster, or perhaps I should say *disasters*. There will be political chaos, turbulence, an unraveling of the current political order. As we are currently, we are unable to cope with the new realities. He is telling me very directly that consciousness must become a focal concern of humans in order for us to proceed further.

"He is right now tapping into *your* [my monitor's] mind. It is like he is locating you, and perhaps measuring or doing something. He is telling me that you are very important in all of this. We must come back here—their world—at later dates. We will be the initial representatives of humans as determined by consciousness. He is telling me that consciousness determined our arrival at this point. There is more. We are not saviors, just initial representatives. He wants me to get this straight.

"I am getting the sense that he wants us to understand that we have a responsibility to represent fairly. This is not to go to our heads. This is just our job now, and we all have jobs. He is also telling me that I am doing a fairly good job at writing all of this down.

"He likes your sense of humor. He says that there will be lots of activity in the future, of the day-to-day sort. But for now, we are to focus on the book. The book is important, and they will use it."

MONITOR: "Tell him 'Thank you.' We need to go now."

CB: "I told him. He already sensed that it was time for me to break off."

MONITOR: "Write down the ending time, Courtney."

CB: *Long pause.* "You might as well tell me the target now."

MONITOR: *Gives a slight nervous chuckle.* "The Federation."

CB: "I see."

DISCUSSION

The implications of this session range from practical aspects to the sublimely philosophical, and readers will want to make up their own minds regarding my interpretations.

There is a galactic governmental organization. I do not know how hierarchical or centralized is its power, nor do I know how many species, or planetary cultures, are represented in this organization. Moreover, I do not know if some groups or cultures have chosen not to join the organization, or if some have been denied membership. My clear sense is, however, that Earth-based humans are being prepared for full membership. It seems that my initial entrance into the council chambers has been interpreted in a positive light in this regard. It may be that one of the rites of passage to membership is for members of a culture to consciously search for the organization via the appropriate means. Indeed, based on what I was told, humans may already be represented in the Federation in a minimal fashion, although I doubt either my monitor or I feel particularly comfortable representing anyone other than ourselves.

Members of the Federation have very advanced levels of consciousness. That is, they completely understand consciousness—both its physical and subspace aspects. Moreover, Federation members seem to think that a widespread enhancement in human understanding of consciousness is a necessary precursor to human participation in galactic life.

I cannot emphasize enough how important the sense of an advanced understanding of consciousness was during the current session. It may be that my own growth in consciousness had to be sufficiently advanced for me to find and enter the chamber, but I have no objective measure of how much this may be true. But it is clear to me that the general concept of growth in consciousness is not a distant goal that the Federation folks think we should pursue. The need for widespread growth in this regard is essential and urgent. Moreover, it seems that humans must begin to examine consciousness from a more practical and scientific perspective, rather than through a distorted bifocal intellectual lens that separates our understandings of physical and nonphysical realities. In my view, until we grow out of our myopia in this regard, we will likely remain a relatively primitive society, and from a planetary-cultural perspective, a galactic backwater.

The UFO abduction literature is filled with reports of interactions with ETs and humans in which the subject of conversation is future planetary disasters on Earth. The causes of the disasters are usually of an ecological and nuclear orientation, and the frequency with which these warnings reappear certainly can give one pause. This session gave me my first indication from a direct source that such planetary problems may indeed occur. At this point in my research, however, it was not clear how these problems were connected to the concepts of humans moving off of the planet.

At the end of this session, the primary thing that my monitor and I agreed upon was that we needed more data—lots of it—in order to comprehend what was turning out to be an increasingly complex situation. We marveled at how naive we both were when we started this research. The idea of simply identifying the ETs who flew the saucers now seemed much too narrow.

CHAPTER 11

The Grey Mind

In order to understand how Greys think, we decided to remote-view the consciousness of the Greys, which would entail my entering the mind of at least one of them. To prepare for a monitored session (Type 4), I decided first to explore the idea of a Grey collective mentality on my own. Remember that with Type 4 data, the monitor does not tell the viewer what the target is until after the session is over. In each instance, it can be a randomly chosen target from a long list upon which we have agreed, or it can be another target chosen by my monitor of which I have heard nothing.

However, with Type 1 data, I am working alone, and I *do* know the target in advance. When working under solo and front-loaded conditions associated with Type 1 data, it is essential to strictly adhere to the structure of the SRV protocols. This places limits on what I can do once I locate the target.

In this chapter I report the results of two sessions—the one in which I alone remote-viewed the target "Grey/massmind," and the other a monitored session using a nearly identical target. I present the material from the solo session in the form of an internal dialogue.

Date: 27 November 1993
Place: Atlanta, Georgia

Data: Type 1
Target coordinates: 7119/5108

_____

Fifteen minutes of preliminaries indicated a sense of movement and energy.

I am picking up grey and white colors. In terms of textures, I am discerning polished, steellike surfaces. It feels very warm here. I hear a strange chirping sound.

In terms of dimensions, I am getting the sense of something very wide and/or open. It seems endless, or perhaps expansive, universal, or just plain huge. There is a strong sense that this . . . whatever . . . goes on and on. It is not a limited or bounded thing.

I am also detecting movement. It is like things and/or energy is moving in and out of a central location.

Now I am picking up something that needs translation. It is different from that which I understand normally. Minimally, it seems like love and caring, but much more all-encompassing than the typical human idea of such things. There is the sense of a caretaker, a mother, or a holder of something precious. In some way, this is a home, a dimensionless home with a sense of infinite freedom.

Moving on, I am also detecting a concern for safety. Something is not right here. There is fear here, and it is strong. Something is stifling. I am picking up the movements of ships, very advanced ships. Again, I sense the fear and the concern for the safety of others.

I get the sense that the Greys are stuck. It is like a birth in which the baby gets stuck in the birth canal. The fear is associated with this condition of stuckness.

Yet surrounding this sense of fear is an envelope of calmness.

DISCUSSION

The collective consciousness of the Greys both protects and nurtures. At the same time, there is a sense of fear within, a deeply emotional flavor that is related to the idea of being trapped. It is as if Greys are trying to emerge from some state, yet the emergence does not work. The calmness that surrounds the fear somehow stabilizes the collective intellect (enabling physical survival) and permits a less terrifying daily existence. The sense of love and protection from within the mind is nearly overpowering from a human perspective. My own personal reaction was one of sadness, and perhaps compassion.

Two and a half months after this session, my monitor decided to have me work the target blind. As readers will observe at the end of the session, he changed a nuance of the target, thereby encouraging a somewhat different approach to the idea. After making the connection with our speakerphones, we chatted in a normal fashion (talking about everything from a new joke that my monitor had heard to the prospects of me lining up a publisher for this book), and then we began.

Date: 11 February 1994
Place: Atlanta, Georgia
Data: Type 4, remotely monitored
Target coordinates: 4384/8296

The preliminaries indicated a sense of liquid and movement.

CB: "OK. I am getting colors like turquoise, and blue. Lots of liquid. Also, rocky and smooth textures. I am getting both warm and cool temperatures as well. I am tasting something salty, and smelling something like fish. Whatever this is, it feels wide and expansive, very deep. I also get energetics."

MONITOR: "Stage Three."

CB: "Drawing now. . . . I just have a horizontal line going across the paper. I am also AOLing on fish and an ocean."

MONITOR: "Write the AOLs down. Stage Four."

CB: "I have liquid. Lots of it. This liquid is a living environment. I get the sense of a birthing place. There is life here. Organisms. This is a place of protection.

"I am getting a sense of a room below the liquid. I am there now, in the room. It seems like a laboratory. Let me put that down as an AOL for now.

"OK. There is a Grey here, in the room. He is staring at me. I am sketching his face now. It seems male, somehow. I am getting the sense that I should go inside his head. What should I do?"

MONITOR: "Let your unconscious solve the problem."

CB: "OK. I am moving in. . . . Wow."

MONITOR: "Keep in structure. Stay in the matrix. What do you have? Write it down."

CB: "There is a sense of emptiness here. Deep emptiness. But at the same time the mind is filled with great awareness, even a feeling of total awareness, whatever that is. This fellow has tasks to do. He is a worker, and he is quite busy. I do not get much of a sense of surface emotion like humans have. If I would compare it to anything, it feels like Focus 15 at the Monroe Institute."

MONITOR: "Can you get a sense of the being's superior?"

CB: "The mind itself is the superior. It is a collective mentality. The collective mind is the controller. No one individual is the superior. All Greys are of the same mind. They are one and together."

MONITOR: "Is there a sense of purpose?"

CB: "Survival is primary, and evolution. It is one collective organism, and survival is paramount, as with any organism. There is a definite lack of distinction between individuals."

MONITOR: "Go to Stage Six. Do a timeline. Mark the timeline where you are now with this being. Now, place the position of the beginning of this session. Probe the timeline for important points. Mark them down, and move on to the Stage Six matrix."

CB: "Doing that now. . . ." *Pause.* "I am now in the matrix. I am getting the sense that there is a collective of many Greys right now. Something is coming across very strongly now. I get the sense that they need to get out of their physical bodies. It is a desperate need. This has a life-and-death quality to it. It is their absolute highest priority for the survival of the mind . . . the organism. The mind is locked up now. It absolutely *must* escape. Coming from very deep, there is near panic in the collective, but in a Grey sense rather than as we would know it.

"The Greys are working with humans and others for a collective escape. It is like getting out of a bottle, or off of a sinking ship. There is a definite sense of panic."

MONITOR: "What would be their ideal environment?"

CB: "They will have a physical planet. Not Earth. They have the ability to terraform a planet and to travel anywhere. They will not push humans off Earth. The universe is too big for that."

MONITOR: "What about their interaction with the Federation?"

CB: "The Greys are members in good standing in the Federation. They participate in many projects and work on many ships. In a Grey sense, they are proud of their abilities to interact with many Federation species."

MONITOR: "Can humans help bootstrap the evolution of the Greys?"

CB: "Not technically. Genetics is a help, and necessary. I also perceive that there are other ways humans will help, but the Greys are not aware of it yet."

MONITOR: "How do Greys experience the idea of leisure?"

CB: "They do not have a human understanding of leisure. To Greys, all time is a continuum. Leisure reflects a need to rest. The Greys do something different."

MONITOR: "What is their lifespan in human terms?"

CB: "Greys are very aware that they do not die. Their physical bodies are seen as clothing or shells, and death is not a meaningful concept to them in the way that we think about it."

MONITOR: "What is the lifespan of typical Grey bodies?"

CB: "Longer than a human lifespan, but it depends. Perhaps two hundred Earth years would be an average."

MONITOR: "Cue again on their ideal environment."

CB: "An ideal environment for the Greys has multiple facets. It is not just a physical planet, since they could get that now. The ideal environment is an evolutionarily new set of individual bodies. The Greys are in a birthing process. They are preparing to leave the collective identity and become linked individuals instead.

"I am going deeper now. . . . This is interesting. At their core, there is a fear and awe, or perhaps wonder at how humans do it—exist and prosper, that is, as individualized identities. They know that they need some of this but are frightened of it as well."

MONITOR: "Cue on the time for physical meetings with Greys and human representatives."

CB: "Many interactions are happening now, but none with human representatives as aware and equal participants."

MONITOR: "What are the protocols for such a meeting, including place and requirements?"

CB: "Even the Greys are confused about this one. They know that a meeting is necessary, but they do not know how to pull one off. They are afraid, in their own way, of interacting with unleashed humans. They do not want to give up the sense of control, or authority. But they know that they must do this somehow. They need help. They are stuck.

"They just asked me if I have any ideas to help. What do I tell them?"

MONITOR: "Tell them we will work on it."

CB: "OK. I just did that. They seem appreciative, in their own way."

MONITOR: "Ask if there should be any involvement with the Martians to help this along."

CB: "The Greys are taking care of the Martians as well. There cannot be much help from them. They have their own problems. They are sick and have little time and few resources to spare for this project.

"I just suggested meeting with the Greys in an uncontrolled environment, and the suggestion was rejected."

MONITOR: "Ask whether they would be willing to meet with you and me, as volunteers, under circumstances that they controlled."

CB: "They accept the idea. Enthusiastically. They tell me that they will work on it right away."

MONITOR: "OK, Courtney. It is time to break away. Write down your ending time."

CB: "Whew! That was something. I am going to need a breather after that one. Well, you can tell me the target now."

MONITOR: "It was 'Greys/consciousness.' "

CB: "It doesn't surprise me."

MONITOR: "Yeah. We need to think this one through."

CB: "Let me take a break and call you back. I will fax you the data immediately. Talk to you in a bit."

DISCUSSION

Although I do not understand why, it appears that the planet that I landed on in the beginning of this monitored session is important to Grey society. Indeed, in some respects—as I discussed later in my research—it resembles the Grey homeworld, especially in terms of its oceans. But I do not know if it is indeed that planet. Nonetheless, it seems that this particular planet, for whatever reasons, has some crucial meaning to these beings. I also sense that the planet is somehow connected to their consciousness, but I do not know how this actually operates.

Based on my own remote-viewing probes, as well as on repeated reports in the UFO abduction literature, Greys communicate telepathically. The sense of a mass mind, or a collective consciousness, is frequently associated with groups of Greys. This idea is difficult to understand in comparison to our own waking-state consciousness. We must improve our understanding of the Greys' mentality if we are ever to interact with them successfully. If the Greys are indeed engaged in a genetic engineering program involving both humans and their own species, it may be that *they need human help* during this particularly difficult time in their own evolution. We must keep an open mind and should not prejudge anything in this regard.

Just as clearly as I can tell that Grey consciousness is a collective mentality, I can see that they need to evolve away from this state. The Grey collective intellect has many positive aspects. Apparently, individual members of their community do not compete with each other in a Darwinian evolutionary battle for supremacy. They either sink or swim together. This is connected to an aspect of altruism in their consciousness that humans might do well to study.

Given the scope and the substance of Grey interactions with humans as reported in the abduction literature (see especially Jacobs 1992; and Mack 1994), it is easy to see how the activities and intentions of Greys could be misunderstood as hostile from a human perspective. But it may turn out that the Greys are less of an invading army than a rescuing cavalry. They may do things to us that we do not understand or like, but they are not evil—of this I am now certain. We simply do not yet understand this species.

Nothing in my experience as a remote viewer suggests that Greys see us as an enemy. They may be afraid both of us and of what we represent, but they need us, spiritually as well as physically. And lest we judge too quickly, we may need them for our own evolutionary survival just as dearly.

In short, we need to know more, much more, about them, ourselves, and the role that we should play in this intensely interesting drama. I suggest that it may be wise to postpone our judgments of all nonhuman species until we advance a bit more in our own understanding of the broader galactic community.

CHAPTER 12

The Human Repository

Both my monitor and I have heard many reports by abductees claiming that at least some ETs are from a world in the Pleiades star cluster. We have never been sure of exactly where this information came from, if indeed it was something other than an unsubstantiated rumor. Neither of us knew whether to consider any of these reports as credible. Nonetheless, and mostly as a test, we decided to remote-view the Pleiades star system in search of sentient life. As it turned out, it was very much worth the effort.

Date: 10 March 1994
Place: Atlanta, Georgia
Data: Type 4, remotely monitored
Target coordinates: 2805/2070

The preliminaries indicated strong energetics and hard, artificial structures.

CB: "I am getting a lot of very bright light. Bright white and yellow. Lots of energetics as well. The temperatures seem to include hot and fiery. I taste something burning, and smell smoke. I am also hearing some crying. Wherever I am, it feels massive, airy, with lots of energetics, radiant, and round. Somehow, this feels important, and weird."

MONITOR: "Stage Three."

CB: "I am sketching a horizon, something burning on the ground, and something very bright in the sky."

MONITOR: "Move to Stage Four."

CB: "OK. I am in the matrix. I am still getting that sense of light, burning, and high temperatures. Whatever it is that is round and fiery is in the sky. I am on the ground now. I am detecting two types of beings. One type is on the ground, and the other type is in something in the air, perhaps a vehicle.

"There is intense work activity going on now. The beings in the air, near the light, are more advanced than the ones on the ground. That light is bright! I can't figure out what the beings are doing near it. It may be that they are in some vehicle near it, but every time I look up I am struck by how bright the light is.

"Focusing on the ground for now, there is dirt, grass, and humans wearing normal American-type casual clothing. Let me check that out. Yep. Pants, socks, shoes, the works. The humans are quite upset. There is a lot of fear here, plus crying. It seems that the humans are quite dazzled by and afraid of the light object or thing in the sky.

"Still with the humans on the ground. They seem to be a family unit, a man, woman, and child. The child is the one crying.

"Looking up now and following the other beings. . . . These beings are very advanced. They are in an object, not in the light thing. I am moving inside now. The object has a round interior. I see the beings now and am moving in close to take a look. These folks seem like Greys."

MONITOR: "Courtney, go to Stage Six. Make a timeline. Place the target time on the line, then place the present time." *Pause.* "Now place the year 2000." *I place this year slightly later than the target time.* "Next, go to the matrix. Now, cue for any significant near past events relating to humankind."

CB: "It seems that humans migrated here."

MONITOR: "What about the crying?"

CB: "The crying may be in response to the light object's appearance. At least the humans are looking up while they are being upset. It could be that they are upset about the vehicle with the Greys, however. Whatever it is that they are upset about, it has to do with something in the sky."

MONITOR: "Describe the people."

CB: "Physically, they have white skin. They seem like farming people, but not primitive. They are comfortable, and perhaps they live in, or visit, a nearby city or village. You know, they really seem like Americans."

MONITOR: "Go back to what is in the sky."

CB: "There is either one very bright vehicle in the sky, which I do not think is correct, or two separate objects, one very bright and the other a vehicle. It is hard to tell, since the bright thing dominates the perspective so much. Whichever it is, the vehicle is over the people on the land, and I sense that the beings in the object wish no harm to them."

MONITOR: "Cue on the intent of the beings in the vehicle."

CB: "The beings in the object are here on a mission. This is a regular job call for them, somewhat routine. Should I go back inside the ship?"

MONITOR: "Instead, cue on the concept 'other places and times visiting.' "

CB: "Doing that now. . . . Wow! I just got whipped across a large distance. It felt like being jerked on a string. I am now about a thousand miles off the surface of an Earth-type planet. There are clouds, water, oceans, and land. My, it is a beautiful planet! The original target and the second seem to be separated by time, space, or both, and perhaps a lot."

MONITOR: "On your Stage Six diagram, put a small circle on the paper for the location of this new planet. Then put a circle on the paper where it feels correct for the location of the original target."

He pauses while I do this. "Now with your pen, probe the circle for the new planet and tell me how many suns it has."

CB: "There is just one."

MONITOR: "OK. Now probe the circle for the original target and tell me how many suns *it* has."

CB: "Gosh. There are two, a large yellow sun and a smaller white dwarf. How can that be?"

MONITOR: "We do not analyze. Stay in structure. Courtney, go back to your Stage Three diagram in which you had the bright object in the sky. Probe the object and dump the data in your Stage Six matrix."

CB: "Hold on. . . . If I look up, it appears to be a fireball. It is emitting lots of energy. It *is* bright, brighter than our sun."

MONITOR: "Let's take a break. Fax me your data up to this point and then call me back."

CB: "OK. Call you in a few minutes."

Resuming

MONITOR: "Courtney, I want you to get more information about the people on the land at the original site."

CB: "Going there now. . . . I am with the people again. They cannot perceive my presence. It is a family: a man, woman, and child. Again, the child is crying. They have hair. They look and feel entirely human, and again, even American. The man has a beard, a large woolly beard. The woman has blond hair. The clothing is colorful. The temperature around here is very comfortable, warm. The man is very confused right now. He really does not understand what is going on."

MONITOR: "OK. Now cue on the beings in the object."

CB: "These are Greys. They know about the humans, of course. Hmmm. This is weird. The Greys know that I am here. They have their attention on me right now. In fact, they are attempting to feed me information very fast. It seems too fast for me to handle.

"The Greys have become a bit confused now, like something else is occupying their attention. There seems to be a bit of a disagreement among them. It could be they are trying to figure out how to feed me information.

"Ah. Things seem to have smoothed out. The people on this world are from Earth. The Greys brought them here. They have been transplanted. The humans do not know everything. They do not even know where they are."

MONITOR: "What is the reason for the transplanting?"

CB: "Human survival is at stake. A new location is needed away from Earth's climatic disasters."

MONITOR: "Continue your probes. Find out more."

CB: "The transplanting is still going on at the target time, but it is not yet happening at our current time. At the present, there are only preparations. They are getting a class M planet ready while waiting for humans to self-destruct."

MONITOR: "What else is being transplanted?"

CB: "Genetic material is dominant. They need as wide a selection as possible of genetic material to ensure the survivability of a better, more advanced gene pool."

MONITOR: "Cue on the needed genetic changes."

CB: "There needs to be a better connection between spirit and body. The current genes downplay this. The original gene structure was necessary for survival in our past. But new or modified genes are needed for growth and survival at this later stage."

MONITOR: "OK, Courtney, let us end here. Put down the ending time."

CB: "Done. Well, what in heaven's name was that?"

MONITOR: "Actually, it was the 'Pleiades star system/cultures.' "

CB: *There is a very long pause in our conversation.* "You're not kidding?"

MONITOR: "Honest. That was it."

CB: "Do you realize what this means?"

MONITOR: "Courtney, I have been in this business so long that nothing surprises me anymore. But this *is* really something."

DISCUSSION

The target for this session led us to a class M planet orbiting a binary star system, around A.D. 2000. Before this session, I had doubts as to whether life could exist within the radiation parameters of a two-sun system. It is possible the Greys have placed some sort of environmental protection on or near the planet to ensure its habitability.

In the near future, there will be humans on this planet who will have been transported there by the Greys from Earth. I do not know if other Earth humans will be aware of this when or after it occurs. It may be done quietly, in which case all information about this new world and its inhabitants will have to come through remote viewing. At target time, the humans are in family groups, they are frightened and confused, and they do not know where they are.

The purpose of the transplantation is to preserve the genetic stock of humans, in the sense of maintaining as wide a gene pool as possible. Some genetic manipulation will be needed later to enhance the mind-body connection for a future breed of humans. This seems to be due to our current destructive tendencies, which are driving the global population in the direction of planetary-wide ecological and climatic disasters. Thus, it seems that some humans will be "space-lifted" to a safe haven while the rest of humanity slugs it out back home. (THE RAPTURE ??)

This was the first time I received information directly from a member of the Greys regarding future planetary problems for humans, but I still do not know the timeline for these impending disasters. My monitor has told me of other remote-viewing data that suggest that Greys are also engaged in collecting plant and animal specimens from Earth at this time. If this is true, one possible reason could be biological needs in terraforming the planet in the Pleiades system.

I personally wonder how many species in our galaxy have the

benefit of such interstellar midwives as the Greys in their evolution. More importantly, what will happen to those humans who remain on Earth? Will they simply perish, or will they continue to evolve in a fashion that is different from their then-separated cousins in the Pleiades?

CHAPTER 13

Reality Check #1

So far, I have presented the results of ten remote-viewing sessions. (There were two in the chapter on the Grey mind.) I have introduced so many new concepts and findings that it would be only natural for the reader to wonder if this could all be a fantasy of mine. Remote viewers wonder too. For an answer to this, we use a target we call a "calibration target." It is something that can be easily verified, and it serves primarily as a check on the use of the SRV protocols.

My monitor and I decided to introduce a few calibration targets into this analysis. The current session contains Type 4 data, meaning I was given *no* information at all about the target either prior to the session or during the session. The session was monitored remotely, and it took approximately twenty-five minutes from start to finish.

Date: 1 May 1994
Place: Atlanta, Georgia
Data: Type 4, remotely monitored
Target coordinates: 4933/4876

The preliminaries indicated a complex man-made structure.

CB: "I am getting browns and tans. The textures are rough, like cement, gritty. It is warm. I am getting something massive, heavy, weighty, even voluminous. I am AOLing a city, so let me put that down."

MONITOR: "Stage Three."

CB: "OK. I have a diagram that looks like a city skyline. There is a main structure in the center and two others on either side."

MONITOR: "OK. Find the target structure and do a movement exercise inside it."

CB: *Long pause.* "I am now getting colors that are grey, black, and white. The textures are polished, smooth, and shiny. The temperatures are comfortably warm. I hear a whirring or buzzing sound. For dimensions, I am getting short, wide, narrow, even squat."

MONITOR: "Stage Three."

CB: "I just have a square on my new Stage Three sketch. It could be anything from a room to a box."

MONITOR: "Stage Four."

CB: "Going through the matrix now. I am getting things that are paperlike. There is something flat and horizontal. This seems like an office."

MONITOR: "Cue on the idea of 'activity.' "

CB: "OK. There are humans walking around here. They have clothes on, business clothes. There are suits, jackets, slacks, and both men and women. They are all wearing business attire. There is a desk in the room. It seems to have something like a blotter on it.

"There is a man sitting at the desk. Wow! This is an important man."

MONITOR: "Put that 'Wow!' down as an AI and move on."

CB: "Maybe I better AOL this. I am staring at President Clinton, straight in the face, as he sits at his desk."

MONITOR: "Stop the session. The target was 'The Oval Office/White House, Washington, D.C.' "

CB: *A low chuckle from both sides of the phone connection.* "Heh. So much for maintaining national security secrets."

MONITOR: "I could have had you go into his head at this point, but that would have been an invasion of privacy. Besides, we accomplished the goal of the session."

CB: "Again, so much for national security secrets. I will fax you this stuff right away."

MONITOR: "Great. Perfect session, Courtney. Take care."

DISCUSSION

The implications of the current session to issues surrounding the maintenance of our national secrets is obvious. Yet there is another issue. It should now be apparent why the ETs know so much more about us than we know about them.

It is hard to imagine anything that would be as disturbing to our governmental agencies as the fact that all of our current secrecy apparatus is now obsolete. I suspect that the ETs *want* us to be disturbed in this respect. By disturbing us, they focus our attention on something that is important to our own growth. If this is indeed their plan, I can only observe that it appears clever in its design to an extreme. I wonder, perhaps with hope, about their own estimates of the probability of its success.

CHAPTER 14

A Diplomatic Breakthrough

One day my monitor reported to me that a remote viewer was recently able to remote-view her own abduction. (She was apparently abducted in classical UFO-ET style many years previously.) The session to which my monitor referred was conducted under highly controlled Type 4 conditions (viewer blind, monitor front-loaded). What made the session's data so surprising was that the viewer was able to view a UFO abduction at all. (Again, this, in general, had not been possible previously; a false replacement signal had always been received.) Neither my monitor nor I drew any conclusions regarding this episode at the time, but we did notice the change. A few days after this conversation, we found out what occurred to make this possible.

Date: 31 May 1994
Place: Atlanta, Georgia
Data: Type 4, remotely monitored
Target coordinates: 3701/5475

The preliminaries indicated a target associated with dry land and a man-made structure.

CB: "I am getting brown and tan colors, rough and woody textures. The temperature is warm and comfortable. I am smelling something like turpentine, and it seems like an outdoors, woodsy smell. I will AOL this as a forest for now."

MONITOR: "Move to Stage Three and then on to Stage Four."

CB: "I am getting a wooden structure, like a house. Moving into the structure. There is a wooden table. This seems a bit oldish. The internal structure of the place is square, or minimally rectangular. This is a very simple dwelling unit."

MONITOR: "Move on to Stage Six. Let's do a movement exercise." *He moves me to 500 feet above the structure.*

CB: "There is a mountain and road near the structure. There is also a river, and apparently either a rapids or a waterfall near the structure as well. Do you want me to investigate this?" *He has me do a variety of procedures designed to identify the location of the structure with regard to nearby topographical features.*

MONITOR: "Courtney, go back into Stage Four and move back inside the structure. Let the unconscious solve the problem."

CB: "OK, I am back inside now. There is a person here, a human male. He is wearing work clothes, a plaid shirt, jeans. He has a beard. I do not pick up much activity of any sort."

MONITOR: "Give yourself some freedom of movement in time. Let your unconscious solve this."

CB: "Give me a moment. OK, I am getting a strong sense of fear from inside the house. It is at night; I have experienced a small time dislocation. I am getting a sense of ET activity—this is strange—an abduction. I am putting this down as an AOL but from the signal line."

MONITOR: "Keep in structure. Continue."

CB: "The table is still there, as is the wood floor. But there are a number of Greys here now. Some of them seem to be floating in the air. They are wearing uniforms, and they are moving quickly, like they have an agenda that needs to be accomplished with great haste.

"There is a woman here now. She seems to be in the center of their activity. They are levitating the woman up and through the window of this house. I do not think that the window is open. She went right through the thing! I am following her and the Greys outside.

"There is a very bright light here. Looking around now. The Greys are taking the woman up into a large ET ship. I am sketching the ship now. I am also sketching the scene inside the house when she was moved through the window."

MONITOR: "Move into the mind of one of the Greys. Find out what they are doing."

CB: "Hold on. Doing that now. This is a survival project. It is their work, their job, in the sense that this is what they do for a living. The idea of survival is very broad, as it is for everyone."

MONITOR: "Now go into the mind of the woman. Stay in the Stage Four matrix."

CB: "There are definitely multiple levels to her understanding of the situation. On the top surface, she is sedated. Just below, but still at the human level, she is terrified. Further below, at the subspace level, she is happy, even ecstatic."

MONITOR: "What is the criterion for selection?"

CB: "She selected herself. She volunteered."

MONITOR: "Go into the ship. Stay in Stage Four."

CB: "Doing that now. Wow!"

MONITOR: "Dump the aesthetic impression in the matrix and move on."

CB: "There are many Greys here. There are huge numbers of operating tables. This is an expansive place, with lots of activity. All of the Greys are very busy. There are other humans on board, apparently all brought on by the Greys."

MONITOR: "Are there any other types of ETs on the ship?"

CB: "There are only Greys and humans on this ship. The woman from the wooden house is on a table now. She is screaming. There is a tall Grey looking between her legs. They are examining her and doing something.

"Apparently I am being noticed. Someone is literally pushing me to look at certain things. It is the sense of being pushed hard into a passage or through a door. I get the sense that someone wants me to see something.

"I am actually in the uterus of the woman now. There is a fetus here. I get the sense of a light, perhaps artificial, lighting up the place. The fetus is on the way out, and I am following the removal of the fetus. The fetus is now outside the woman. She is very calm, exhausted, perhaps just 'out of it.' She may have fainted. The fetus is being put quickly into a liquid in a clear canister."

MONITOR: "How long is the fetus in the canister?"

CB: "A short time for this particular tube. I am now getting this information from an adult Grey who is patiently standing near me. The being seems to be some kind of nurse or midwife—a caretaker personality. The Grey is telling me that a baby gets shuffled about as it matures. Later it goes into a larger canister until it comes out completely. Ultimately, the removal is like a normal birth. The baby is taken out and given air to breathe."

MONITOR: "How long has this operation been going on?"

CB: "I am being told that this is a major breakthrough with humans. The decision has been made just recently to show humans [i.e., us] the entire project. The Greys will no longer interfere with our remote-viewing attempts. We can now watch the so-called abductions freely. They are hoping for a change in Grey-human relations, and they are doing this as a major concession [although, perhaps this is the wrong word] on their part. They want to work cooperatively with humans.

"With regard to the current intensity of their activities, this is a new operation. The project has taken on a new direction, from continuity and slow change of the past to rapid, major evolutionary advances that will allow Greys and humans to be self-determining.

"Apparently this change has been dictated by larger, Federation decisions. Change is required of the Greys, and they also are required to help humans. There is a definite sense of long-term appreciation of help regarding what the humans have given—voluntarily—to the Greys."

MONITOR: "OK, that is enough for now. End the session."

CB: "That session took a long time. I'm exhausted. Well, tell me what the target was."

MONITOR: " 'Federation/current Earth operations.' "

CB: "Hmmm. Looks like things are going to be different from now on."

MONITOR: "It gives one pause."

DISCUSSION

It should be noted that the current session took place only a few weeks after the release of John Mack's book on the ET abduction phenomenon. Mack's treatment of the phenomenon is perhaps the most sensitive—and positive with respect to the Greys—in the extant literature. It could be that the Greys decided to change their tactics regarding our remote-viewing attempts because they concluded that we are now capable of understanding the phenomenon without the fear that was so typical of our previous reactions to their activities. However, it could also be that they decided that hiding their activities from human remote viewers served little purpose given the fact that so much information had already been uncovered using hypnosis as the primary investigative tool. Yet the additional possibility exists that the Greys will now allow us to see their activities because of the positive nature of Mack's book, perhaps thinking that they might now get a fair shake in the inevitable trial that will emerge in the court of public opinion.

Following this session, my monitor interacted with a number of remote viewers who targeted ET abductions under highly controlled Type 4 settings, and all met significant success. Sometime during the month of May 1994, the Greys had changed their behavior with regard to how they chose to interact with humans, a highly significant act because it demonstrates an evolutionary change in how Greys think about humans. We are being considered on a somewhat more equal level with them, at least in the sense that we can watch each other's activities freely. I suspect that this could be seen as a precursor to much greater cooperation in the future.

Perhaps the most important datum to come from my current session was that at least one abductee was abducted voluntarily. The idea that I got during the session was that this particular abductee had multiple layers of awareness within a complex structure of consciousness. At a deep level, she was aware of the abduction experience and was happy to participate in it. Yet her conscious mind neither remembered any previous agreement to participate, nor did it want to participate further in that which seemed only horrific.

Given the multilayered structure of human consciousness, I can also see how confusing it would be for the Greys and why they might have conducted the abductions in a seemingly callous manner. They would be aware of the person's original agreement to participate (perhaps made before physical birth), and would thus disregard the physical human's panic. Perhaps they would assume that all will be forgiven when the abductee's physical body dies and its subspace personality becomes freed from the confounding influence of physical awareness.

It seems recent human actions have encouraged a change in Grey behavior. If the Greys are indeed working to help humans as well as themselves, they may have much to gain by treating us as active partners in this venture.

CHAPTER 15

Jesus

As research into this project progressed, it became obvious to me that the implications of these findings for virtually all of humanity would be enormous. Thus, I decided to add a few new targets to our longer list of targets. These targets were individuals who had been instrumental in guiding humanity at critical moments during our history. These were teachers to whom many came for advice. Jesus was one such personality that I added to the list.

At first my monitor was hesitant to add Jesus to the list. I suppose he had a few reasons. I would be hitting the target blind, in the sense that I would be given only the coordinate numbers for the session, and one might think that this may not be a particularly reverent way to approach Jesus. Perhaps our greatest fear was that Jesus would not want to participate in our study and that he would ignore our request for a conference. Honestly, we did not know what would happen.

To add to the complexity of the problem, we were expanding the purpose of SRV as it was originally structured. The original purpose of SRV was to extract descriptive information about a physical site through the passive act of observation. But by targeting a specific personality, we were consciously attempting to communicate with a known sentient entity. There was no reason to suspect that this could not be done, since many nonphysical beings had communicated with us once we inadvertently bumped into them while remote viewing. But there had never been an attempt to establish a communicative dialogue as the purpose of the session. Thus, this session helped set the stage for a highly significant advance in the development of SRV. We now know without doubt that SRV can be used as a reliable nonpassive mechanism for communication.

This chapter includes the data for two sessions in which Jesus was the target. Quite honestly, the second session was required because my monitor became so flustered at the end of the first session when I correctly identified the target that he had to break off the session early before having me ask some of the interesting questions that I needed answered. After the session ended and he told me the target identification, I too was a bit taken aback. It took us a while to get over the surprise that Jesus was a friendly personality who did not mind that we approach him for advice under highly controlled Type 4 settings. We were also interested in the fact that Jesus seemed willing to act as a creative teacher within the structure of the SRV protocols, in the sense that he set up a show for us to see before we met him face to face.

Date: 2 June 1994
Place: Atlanta, Georgia
Data: Type 4, remotely monitored
Target coordinates: 8863/8473

The preliminaries initially indicated that the target was associated with a hard, man-made structure.

CB: "I am hearing machinery. The temperatures are warm, and there is a bitter taste and the smell of smoke. I am picking up a lot of energy and activity in a condensed place. I am going to AOL this as a construction site.

"Moving on to the Stage Three sketch. I am drawing what appears to be a site with some structures, smoke, construction vehicles, and a pit."

MONITOR: "Put those down as AOLs and move on to Stage Four."

CB: "OK. I have smoke. Lots of burning; stinky. The place feels congested. There is lots of working activity. As an emotional impact for the site, I am getting the sense of frenetic concern.

"I am getting at least one vehicle now. There are beings here. They have clothes—work clothes, jeans. There are hard hats on the beings. Again, as an EI [emotional impact], I am picking up panic associated with the site. I am again perceiving this as a construction site, and I will declare this as an AOL of the signal line."

MONITOR: "Cue on the concept of construction."

CB: "Hold on. . . . These folks are making something. The construction workers are in a hurry. I am moving forward in time now. . . . I perceive a large constructed and polished building. I am getting a strong AOL now. I will put this down as of the signal line. This seems like the new public health building that is currently being constructed here at Emory University."

MONITOR: "Go inside the building and into the minds of the people that are there."

CB: "There is an urgent health problem, a planetary concern. There seems to be a connection in the minds of these people with the Centers for Disease Control. I am getting an overwhelming AOL. I know this building. This is the new public health building that is being built at Emory right next to the CDC."

MONITOR: "You can put that down as an AOL/matching. Then cue on the health problem."

CB: "Doing that now. . . . There are multiple problems, many places on the planet. There is the combination of starvation and sickness. There are new diseases, new forms of bacteria, new viruses, even radiation mutations. The folks in the public health building are trying to figure out ways to control the situation as it simultaneously is spinning out of control."

MONITOR: "Cue on the concept of guidance/help."

CB: "OK. Hold on. There is a need for humans to go to basics in living. I am getting the clear perception that a quick technological fix is not possible.

"Hold on, something is happening. I am now perceiving that the information of this entire session is coming from some being. I am moving in on the being now. Hmmm. This seems to be a light being. He is somewhat translucent. He is wearing a gown, and his hair seems to be made of light. I am getting the flavor of a spiritual presence. I am getting the sense this guy is Jesus; I'll put that down as an AOL of the signal line. I am also perceiving a good deal of love projected toward me from this fellow.

"This being seems to be telling me that the situation has been set up so that no physical solution can remedy the problem. The idea is to force humans out of their physical entrapment, and thereby save the race."

DISCUSSION

At this point in the session, my monitor's voice changed perceptibly. He seemed a bit nervous and quickly called an end to the session. I asked him what the target was, and after a pause, he said, "That was Jesus."

Still in a somewhat bilocated state, I replied, "Jesus?! You're not kidding? That target really *was* Jesus? What was he doing there?"

My monitor simply told me that this was a momentous event in his life and in the development of remote viewing, and that he had to get off the phone and think about its implications.

Before he hung up, I blurted out, "But the guy seemed rather friendly, and I detected a sense of humor. I even got the sense that he expected to see me again this way. I did not even get to ask him about the Martians or the Greys! What are we supposed to do about them?"

My monitor just wanted to get off the phone and think about all of this, so I signed off. After I became unbilocated, I began to realize how momentous this session really was. First, it demonstrated that one could successfully target a person who had once been a physical human and who had died. Second, the session demonstrated unequivocally that SRV could be used to establish communication between two beings (i.e., the viewer and someone else). Third, it became obvious that the targeted individual can sometimes control the flow of information toward the viewer. In this case, Jesus had us travel through the logistics of a planetary health problem before revealing himself. I suspect that the personality of the target is very influential in this setting. Jesus wanted to teach us something experientially, not just with a lecture. Thus, he set up a learning situation for us. Also, I realized that some of the other personalities that we added to the list may decide to employ similar techniques in their communications. I had no idea when my monitor would hit me with these other targets, but I was certain at this point that surprise was to be expected.

The lesson itself is very significant as well. Apparently there is a connection between the health disasters that humans will face in the near future and the activities of the extraterrestrials on Earth. At this point in the project, it seemed that a complex theatrical undertaking involving many aspects and groups was being orchestrated. The basic idea of the phenomenon is to force humans to transform their attitudes and behaviors in various ways while at the same time exploiting this opportunity of transformation to introduce humans to the broader galactic community, and, indeed, to their new responsibilities as maturing galactic citizens. But these ideas were mere speculation at this point. I needed to know more about the broader picture before I would be able to make definitive conclusions.

Approximately two weeks after the original, monitored session targeting Jesus, I decided to target Jesus a second time in order to establish a dialogue, this time under Type 1 (solo and front-loaded) data conditions. At this point I was becoming quite proficient with the use of the SRV protocols, and my ability to obtain accurate Type 1 data was not in question. As with other Type 1 data sessions, I present the information in the form of a narrative.

Date: 14 June 1994
Place: Atlanta, Georgia
Data: Type 1

My initial impressions were of blue, white, and yellow colors. The texture was airy. I got the sense of something expansive and curving. In later stages of the session, I picked up expansive energetics together with a sense of calmness. I got the sense of a resting place. The initial image was one of a large circular body with a luminous and energy-radiating atmosphere. I followed the signal into the light.

Soon after this, I began to discern beings, and I got the clear sense that the beings were waiting for me. Again, I got the sense of calmness from the surroundings. I felt I was in some sort of protection zone, a place to recover (from what, I did not know).

Continuing to follow the signal, I approached a human-type face. There were five other beings present. They were all wearing white translucent gowns, and I noted that I could see through all of the beings. I located the central figure, and probed on the question of what they wanted me to do. They indicated that they wanted to take me somewhere and that I should follow.

We proceeded into a large room that had translucent walls. There were many other beings in the room. At this point I had a strong AOL that this was the Federation headquarters that I had visited earlier, and I realized that my initial image in this session of the large circular body with a luminous energetic atmosphere was similar to that which I had perceived when approaching the headquarters in a previous session. All of the beings in the current room were wearing the same translucent white gowns that the five original beings wore.

I perceived two strong AOLs at this point. The first was that this place was a central command chamber of some type. The second was that it had the flavor of a military headquarters for command and control of various operations. In the room there were tables and chairs as well.

I was taken face to face to the same heavyset Buddha-type fellow with whom I had interacted previously in my visit to the Federation headquarters. I got the sense that he was in charge of the place in some way. He directed me to enter his mind, and I went in.

Upon entering his mind, I simultaneously entered into another dimension or realm. It was as if one dimension was behind the other. In this other place, I cued on the concept of guidance, since that was the cue that led my unconscious to reveal the Jesus personality in the previous session. I then saw his face. Immediately after perceiving Jesus' face, I got the distinct sense that he was glad that I returned to see him alone (i.e., in a solo session).

To obtain information from Jesus within the structure of the SRV protocols, I cued on the concept of human-Grey interactions. The response that I received was very clear and even authoritative. He said that there is no being that humans will interact with that is not of his design. He then stated that we are to help his children however they come to us.

I did not understand what Jesus meant by "of his design" or "his children." It could be that he was using words that would be understandable to a larger human audience, some of whom look on him as a religious figure. I suspected that his meaning would be more complex than simple or literal. Other readers may interpret his words differently.

Jesus went on to say that we humans have free will, and that we have the capabilities to sort out what to do with our interactions with the Greys. However, I got the clear impression that our choices will determine much of our future. I then cued on the concept of human-Martian interactions and got a response similar to that regarding the Greys.

I then cued on the dual concepts of mind and the Sidhis. Jesus was very clear in his response to this. He said that there are countless ways that the spirit is guided to the source. It is like a river and its tributaries. There is no one way, and the practice of the Sidhis is *not* the only way for humans to evolve. But he added that the Sidhis are nonetheless a useful approach for the human mind. I should add that I got the sense that other types of minds may not need exercises like the Sidhis to discern nonphysical realities. The Sidhis are useful to the human mind, and perhaps only the human mind. I got no sense at all of how broadly it could be applied in other nonhuman settings.

I cued on the concept of dangers and received the response that greed is the killer of personalities. It is not compatible with the remainder of life. It is like oil and water: love and greed do not mix. I did not get the impression that Jesus was moralizing. It was as if he was simply stating a fact of life.

Cuing on the concept of ecology, Jesus stated that God can create and re-create all life. The purpose of life is to produce evolution. Following this, I cued on the idea of the Galactic Federation. Jesus indicated that the beings involved in the Federation are at a higher level of evolution than humans. While they too are working to enhance their own evolution, their actions are no more or less crucial than human actions. Moreover, he emphatically claimed that they do not work for him specifically. They work for themselves, and their own growth. They see their progress as leading toward a God-destiny more clearly than humans, however.

I then asked Jesus why I came to him through the Federation

folks. He told me that it was specifically because of the book that I am writing. He wanted to help me with the book because it is my evolutionary contribution. I got the sense that it is one *small* act that assists everyone else. No one person can go forward unless he or she helps the others who temporarily remain. It is a law of evolution, the opposite is selfishness and greed.

I then asked whether humans should view Greys and Martians with compassion. The answer was yes. This is the idea of helping others. Without this, no one goes forward. There must be *no* limits to the human desire to help. This is the most absolute of *commands* (emphasis his). Prejudice—racial, species, or otherwise—simply cannot co-exist with the higher evolutionary forms. This is the human challenge: to grow past the intellectual and habitual limitations of the past that ultimately restrict our freedom.

At this point, I thanked Jesus and ended the session. In my notes for the session, I emphasized my sense that the overall tone of the session was matter-of-fact.

DISCUSSION

Jesus is much concerned with the evolutionary growth of humans. Moreover, he is willing to participate directly in at least some human projects involving attempts to establish verifiable contact with him.

Because of his historical importance, using SRV I asked him his opinions regarding many of the momentous events of our own time. He offered his opinions freely. Nothing more has been done, and my results are not intended to challenge any existing religious concepts. For example, one does not need to believe that Jesus is the Christian concept of the son of God in order find his views interesting, nor does believing that change anything. He once existed as a physical being on our planet, and his subspace aspect is still alive and well, just as the subspace aspects of all of us will outlive our own physical bodies.

Similarly to this chapter, later chapters will present data involving interactions with the personalities of Buddha and Guru Dev. My approach to these personalities is in part out of respect to the roles that they have played in the development of human culture over the centuries. But also I believe it is wise to ask wise beings for advice.

I will have more to say in later chapters on the advice that Jesus has given us regarding the ETs and the way our own evolution can be impacted by our decisions regarding them. For now it is perhaps sufficient to say that he is not silent on the matter. He wants us to work with the Greys and Martians. He did not tell me that it would simply be a good idea. He conveyed the clearest concept of a command that I have ever received from any being while remote viewing. It is not that we *should* cooperatively work with the ETs. Rather, we *must* do so.

CHAPTER 16

The Cause of the Collapse of the Early Grey Civilization

One of the essential goals of my research has been to understand early Grey civilization. I wanted to know about their homeworld, the major events that have occurred in their history, and the more important challenges that they have faced as a culture.

This chapter presents the data from three separate remote-viewing sessions. The first is a monitored and blind session (Type 4 data) in which the cue was "Greys/early civilization." The second and third sessions were done solo under Type 1 settings. The second session on which I report was actually conducted first chronologically, approximately six months before my monitor finally worked the blind session.

Date: 16 June 1994
Place: Atlanta, Georgia
Data: Type 4, remotely monitored
Target coordinates: 4923/8216

The preliminaries indicated a composite target with dry land, a body of liquid, and artificial structures.

CB: "I am getting colors of blue and black. The textures are wet and splashy. The temperatures are cool, and the tastes are salty. The smells are fishy, and I hear sounds of splashing.

"I have some sort of structure on a flat surface. I am drawing this now on my Stage Three sketch. The surface seems to be wet, like a body of water. The structure is hard." *The monitor gives me a movement exercise that places me on the top of the structure.*

"There is a lot of spiraling energetics below this structure. Like a funnel of some sort." *I draw another Stage 3 sketch of a structure on the surface of a body of water. Below the structure is a large vortex of spiraling water, like a huge whirlpool. The water around the structure (outside of the vortex) is very turbulent.*

MONITOR: "Courtney, move on to Stage Four."

CB: "OK. I'm there now, in the matrix. I definitely have a vortex of some sort. There is plenty of water, and a lot of power involved in this site. There are beings as well. I get the sense that this is a job environment for them.

"Moving in on the vortex, there is tremendous energy connected with a huge volume of water here. The vortex is sucking downward.

"The beings are humanoid, very advanced. I am moving in on the structure now. The thing is some kind of ET thing, perhaps a ship, but not necessarily a spaceship. Shifting back to the beings. Their job is environmental and it deals with oceans. There is something here about fish diversity in connection with the ocean habitat. The idea is of saving/preserving/preparing for some change. There is something here connected with a concern for destructiveness in the habitat.

"I am in the structure now, looking at the inside of the thing. There are computer or control panels here, not so advanced or complicated as other ET spaceships that I have seen. It seems like the inside of this ship is a hospitable and sterile environment for the humanoid beings. There is a clean smell here.

"I am moving in on the beings again. They are definitely humanoid. Getting closer. They are of the Grey type, but something is weird here. The eyes are not so big, but these are definitely Greys. Their eyes are sunken as well.

"These folks seem to have noticed me now. They also seem a bit surprised with my appearance here. This is unusual. I have never surprised a Grey before.

"I get the idea that they are respectful of my visit and have just now been told by superiors—or actually, someone to whom they defer—to cooperate with me."

MONITOR: "What can you tell me about their clothing? Are they wearing anything?"

CB: "They have uniforms on. They even have an insignia on the uniforms. I am sketching it now. Hmmm. It is the same thing that I saw on the uniforms of the Greys who rescued the Martians on Mars, the heart-shaped thing with a snake thing inside. The insignia is theirs. It is the symbol for their unit or corps.

"I am entering their minds now. I get a sort of blank, like my understanding does not exactly match theirs. But it is not so strong a blank as I have had with other Greys."

MONITOR: "See if you can get information from them about where they are."

CB: "Doing that now. . . . The question has locational data, and they are initially concerned about this. They seem disturbed that there could be some confusion, and perhaps worry with this information. I am not pressing it, but I get the sense I might be able to. This is not Earth, however."

My monitor has me execute a movement exercise that places me at the home port of the ship.

"I have a city type of setting, with an open area in front of it from my current perspective. There are buildings here, polished and smooth. Many of the buildings are pointy at the top, like spires. The smell of the place is pungent, sooty. I am getting the strong AOL that this is the Grey homeworld."

MONITOR: "Courtney, from this perspective, probe the meaning of the insignia on the uniforms."

CB: "It represents the symbol of a rescue corps. There is some type of disaster/calamity taking place, and that unit is acting to save something. I am getting the clear sense that the particular unit or corps had its origin in the period of collapse of the early Grey civilization."

MONITOR: "Probe the concept of collapse."

CB: "Hold on. I am getting the smell of burning, pungent, sooty, stinky. There is tremendous environmental waste involved here."

MONITOR: "What about the leadership?"

CB: "This Grey society has a very selfish orientation. The collective society institutionalized the drive for greed. The idea is to accept as normal the individual's taking whatever is desired and not caring for the overall good. I am getting a weird AOL. I am AOLing something like a rebellion in subspace. I am going to put this down as similar to the Lucifer Rebellion.[5] I have to put that down as coming from the signal line."

MONITOR: "Just put it down. Don't try to interpret anything. Move on. Continue to probe the concept of calamity/rescue/recovery."

CB: "The rescue corps is mobilized from within the society. They are the ones who came up with the insignia. Hold on, something is changing. . . . I am now in direct contact with a particular Grey whose job is to help me understand what is happening. This Grey seems like the normal ones, i.e., with the big wraparound eyes. I am getting the sense that this Grey thinks that I have been a bit confused in this session and it is necessary for the Greys to clarify things that are important. I am back viewing the buildings. I am drawing a better sketch of the place now.

"Following the signal, I am moving in on the concept of environmental collapse. There is total pollution here. Literally, these beings are swimming in their own feces. Their entire consciousness is oriented toward self-gratification.

"I am probing the concept of sex now. It seems that these folks are extremely sexually motivated.

"Moving on to food. Their food is mass-produced. There are many individuals to feed, literally billions. Over time, the food became highly processed and very far from a natural design. The source of food was the oceans originally. I get the idea that these folks eat fish.

---

5. Some remote viewers have obtained data that suggest that Lucifer and Satan are (or perhaps were) two real and separate personalities involved in a subspace war that did not turn out well for them. Early mystics apparently perceived some of these events and included references to this war in their writings, which were later incorporated into religious beliefs.

"Again, following the signal. I sense that these beings were corrupted by some type of subspace war. It's as if they were collectively seduced by an arrogant, rebellious, and very powerful leader. They later felt betrayed, but the damage was too far gone. They had to recover from scratch."

MONITOR: "OK, Courtney. Let's end the session."

CB: "And the target was . . . ?"

MONITOR: " 'Greys/early civilization.' "

The data presented above were corroborated by a solo session (Type 1 data) that I conducted and recorded half a year earlier, on December 3, 1993. In that earlier session, I perceived a very congested and highly polluted city. The atmosphere on the Grey homeworld seemed harsh, perhaps caustic by Earth standards, and this harshness was not natural.

With regard to the Greys themselves, I also perceived some of the vocal expressions of the early Greys. Their voices made curious chirping sounds. I also got the impression that the psychology of the Greys of that era had something in common with human psychology. The earlier Greys had smaller eyes than the Greys that are now active on and around Earth. Their skin was light-colored and smooth and it wrinkled when touched. They had no hair that I perceived. In terms of sexuality, the early Greys had a highly focused sex drive, almost single-minded.

In that earlier session, I also saw that the early Greys' babies were in trouble, sick as a consequence of pollution. Many deaths resulted from this. The Greys began having trouble reproducing. They gave birth, but with difficulty. They also did not have a clear idea of how much trouble they were in, in the sense that the early society was quite naive with regard to their dilemma (not unlike humans today).

After the monitored session, I decided to hit the target a third time so that I could fill in some of the important details that seemed to be missing from both of the previous monitored and solo sessions. Thus, under Type 1 conditions, I targeted "Greys/early civilization" on June 20, 1994.

In the beginning of the session, I relocated the structure above the vortex in the ocean that I perceived in the monitored session. More detailed investigation revealed that this was a food-production and maintenance facility. The vortex in the water was related to this activity.

Focusing on the Greys in the structure, I noticed that their genitalia were quite small by human standards. Both males and females worked together in the structure in a seemingly egalitarian fashion. They had robust sexual lives, to say the least. Indeed, the quantity of their sexual activities seemed to dwarf that of humans, both in terms of the frequency of the act and the multiplicity of partners. Telepathy was more advanced among these beings than humans, and the sex act's telepathic component intensified the experience.

In terms of the homeworld itself, I perceived that the atmosphere had been damaged. In addition to pollution, radiation bombarded the Greys at unhealthy levels.

At this point of the solo session, I began probing the cause of the Grey problems on their homeworld. It was clear that there was an evolutionary mistake of some sort. The focus on the self for self-gratification led to a behavioral dysfunction on the part of the vast majority of Greys. There seemed to be nothing that could counterbalance the drive for physical pleasure in their psyches.

Following the signal further, I cued on the idea of spirituality. At this point, the session changed remarkably. I began to perceive a subspace light being. My sense was that this amorphous being was particularly powerful in some way unknown to me. Initially, I perceived both dark and light areas in it, and it did not seem benign.

I then cued on the idea of external agents to the problems of the Greys, and I experienced a locational shift. I found myself overlooking what I can only describe as a layer of subspace life in which there were multitudes of subspace beings. In this layer of life, there was tremendous commotion, like a subspace version of Grand Central Station at rush hour in Manhattan. The magnitude of the commotion and chaos in this layer of life was almost overwhelming. Probing on the relationship between these subspace beings and the Greys, I discerned that these subspace beings *were* the Greys before their birth in physical bodies.

I detected an organizational structure among the subspace beings, and pursuing it, found they had a rigid and hierarchical social order. The control over their existence within this hierarchy was almost military in quality. They took orders and followed them. Strangely, they had been *ordered* to self-indulge and destroy (both in subspace and after physical birth).

Following the signal further, I went after the leadership of the organization. I found myself in a subspace command-and-control center. The center had about ten beings in it at the time. Four or five of the beings seemed to be of a higher authority than the others. The internal aspect of the structure was arranged like an office building, and it became ever more apparent a rigid military structure prevailed here. I continued to follow the signal until I arrived at the one dominant leader of the organization. This being was the same amorphously shaped dark and light being that I had perceived earlier in the session.

I entered its mind only to find it had an extremely dark mind. Something was very wrong there. It was as if the being was psychologically ill.

To begin with, it had a pathological fear of dying. It seemed to think that military fighting and conquest was needed in order to survive. It knew that mistakes had been made, and it was afraid of punishment. The leader seemed unable to devise a plan for reconciliation—fear prevented it. Then it became clear to me: this leader was a terrorist.

Continuing the mind probe, this subspace terrorist leader was intent on destroying the Grey homeworld. The aim was to instill fear in the other parts of the realm, and thus weaken the opposing forces. Fear was the key weapon. In words that best reflect the intent of the leader's mind, the Grey souls were being held as hostages during the crisis. The dark mind wanted a negotiated settlement that would establish its right of personality survival, but with changes. It wanted control over its own dominion. It wanted to establish itself as a sovereign—a dictator.

Indeed, the leader desired worship (of itself). Its need for worship was built into a weakness in its personality structure. It needed worship in order to assist its own flawed personality. In a weird way, the leader had a problem with low self-esteem.

As I was making these observations, I felt the being shift its attention to me. It executed a time and locational shift to find me, then I felt it "descend" into my office, like a dark subspace cloud surrounding me as I sat at my desk.

To my considerable interest, I felt absolutely no fear of this being. I simply examined it, and it examined me. It then departed after a few seconds of observation (perhaps thirty seconds), leaving the impression that it felt that I was a small-fry being that was not directly threatening its activities or reign. In short, I was a pest.

DISCUSSION

Apparently, there was a subspace cause underlying the collapse of the Grey civilization. While I do not completely understand all of the events that led up to this collapse, I have no doubt that the Greys killed their homeworld. They then had to devise strategies to survive. Since their current physical appearance differs from that of their earlier period, I can only assume that their subsequent experiences were responsible for a considerable metamorphosis on their part. The growth in the size of their eyes suggests that they began to live in a darker environment, perhaps underground, which makes sense given that the surface environment of their homeworld was in rapid decay.

The present-day lack of sexual activity on the part of current Greys suggests that they genetically rid themselves of this physiological process. Indeed, it appears as if they genetically castrated themselves. The reason for this could be their need to control their own population size. However, it also implies that they developed alternative means of reproducing, perhaps using technological means to support fetus development. This idea closely corresponds with reports of test-tube babies and fetuses in liquid-filled incubators that consistently appear in the UFO abduction literature (in particular, see Mack 1994 and Jacobs 1992).

Perhaps, though, another reason underlies the Greys' choice to rid themselves of their sexual function. It is not just that the Greys do not reproduce sexually anymore; they also do not have the psychological drive to engage in sexual activity. My own speculation is that their experiences during the collapse of their early civilization made them re-evaluate the underlying causes of their dysfunctionality. They may have felt that their sexual drive helped lead them into their dilemma. The project of gene manipulation—initially intended to quickly adapt themselves to their new and harsher environment—may have extended more broadly into the sexual functioning of their minds as well.

As a final point of interest, I note that some of my observations and analyses find correspondence (not corroboration, but overlap) with ideas raised by Royal and Priest (1992), who base their research on what they claim is channeled data.

Many questions remain with regard to the subject of the collapse of early Grey civilization. I do not know much about the rebel leader who seemed to be orchestrating the collapse. The collapse itself seems to have been an act of terrorism. But against whom was the rebel leader fighting? What mistakes were made that made the leader seek rebellion as a means of survival? It is also unclear as to exactly how the activity in subspace translated itself into the physical world. Humans have a very difficult time perceiving subspace activity. But early Grey physiology may have allowed a more transparent approach between the dimensions, and the corruption of physical life may have been easily accomplished on their homeworld.

Finally, who were the beings who worked directly under the command of the rebel leader? They had humanoid form, but the rebel leader definitely did not. Previous research has indicated that subspace beings have some control over their appearance, whereas physical beings do not. Did the amorphous shape enhance the leader's ability to terrorize? I simply do not know the answers to these questions. But one thing is certain. The collapse of the early Grey civilization was not a simple process. There were both physical and subspace aspects, and for our own survival, we would be wise to learn from their mistakes in both dimensions.

CHAPTER 17

Star Trek and the ET-Assisted Transformation of Human Culture

During the two years I conducted the research for this book, I was often struck by the similarities between many of the ideas that were presented in the television show *Star Trek: The Next Generation* and the data obtained about real ET activities through remote viewing. After the television broadcast of the final episode of this series in the spring of 1994, I asked my monitor to add a new item to our list of targets that would help us resolve the question of ET influence in the generation of the *Star Trek* series.

My original goal was to learn whether ETs were somehow manipulating the minds of the writers so that they would come up with ideas for the show. I assumed that the ETs wanted human culture to become more open to the complexities of galactic life, and popular television shows would be one way that ETs could indirectly mold the collective thinking of the broader public regarding such things. In particular, since *Star Trek* is watched by so many young people, inserting realistic concepts into the show would be an ideal way to educate the next generation of humans along these lines. At least, these were the ideas that I shared with my monitor during our initial discussions regarding the addition of this new target. As it turns out, he had suspected that this type of ET manipulation of Hollywood products had been occurring for a long time, and he recalled that some members of the military remote-viewing team also supported his suspicions.

This chapter presents the results of two remote-viewing sessions. The first is a monitored session in which I was assigned a *Star Trek*–related target under Type 4, blind, conditions. The second is a solo session conducted under Type 1 settings. I conducted the second session to obtain answers to some important questions raised by the first session.

Date: 1 July 1994
Place: Atlanta, Georgia
Data: Type 4, remotely monitored
Target coordinates: 8074/7435

The preliminaries (this time up through the beginning of Stage 4) gave the indication that there were two sites directly involved with the target. The Grey homeworld was one of these sites. My monitor had me direct my initial attention toward the other site.

CB: "I am perceiving the colors of black, grey, and tan. The textures are rough, smooth, some polished, and leafy. The temperatures are warm, and cool in some places. I have a slight taste of salt, and smell some dust."

I sketch a scene with a buildinglike structure and some foliage before moving into Stage 4 to obtain more detailed data.

"OK. I have trees at this site. There is a forest of some type, not too extensive. There is water here as well, like a river or stream. The flow of the river is quite strong.

"There is a structure on this site as well. I am moving inside the structure now. There are lots of people here. Human beings. They are all wearing modern business suits. Hmmm. I am expanding my awareness now because I have noticed that there are also lots of nonphysical beings in this room as well. Wow, there is lots of subspace activity in this room. I sense that the subspace beings are aware of me, but I am not the focus of their concern. They are really busy at work here with these physical people.

"It seems that the subspace beings are not so concerned about the actual humans in this structure, but about the activities that are being conducted by these physical humans. Somehow, this activity relates to some change, both better and worse, regarding the planet Earth."

MONITOR: "How many subspace beings are there?"

CB: "Many. Maybe ten or more. They are humanoid in appearance. They are all wearing white, luminous robes."

My monitor has me move to the other site that is indicated in the preliminary procedures for this session. This site, again, is the Grey homeworld. I experience a time shift in arriving at this site, indicating that the two sites are interacting simultaneously, but from different points in time. Readers should be reminded that in subspace, all time is simultaneous, and it is straightforward for something in the past or future in one location to interact with the present in another location.

In Stage 6, I begin to explore what connected the Grey homeworld with the subspace beings and humans in the structure on Earth.

"The subspace beings in the structure are former human beings. I am seeing that they are closely working with the Greys on an Earth project relating to physical humans.

"From the Grey homeworld, I am perceiving that a tremendous amount of subspace energy is being generated in support of this project. There is a lot of white light relating to this. I have no idea what all of this is about.

"Hold on, the subspace beings in the structure, call it a room, are directing their attention to me now. Things are changing. Someone seems to have told them to fill me in on what is going on. They are really busy, and I get the idea that their activity is difficult right now. They do not want to leave what they are doing, but somehow they have been told that I need to be straightened out first, as a high priority.

"OK, this is what is going on. This is a project. They are telling me that Earth is not going to be useful for anyone to evolve on in the near future if the habitat—widely defined—is not fixed. There is a desperate need to coordinate their activities with the Greys in order to shift human awareness of the problems the planet and human society are facing right now."

MONITOR: "Cue on the concept of human awareness."

CB: "The most important single ingredient involved in the catalytic transformation of human culture is the physical human awareness of subspace life and existence. I am being told that the physical exists to help the evolution of life in subspace, not the reverse.

"I am shifting now to the energetics on the Grey world. It seems that the machinery necessary to work with this quantity of energy is only available (conveniently to Earth) on the Grey homeworld. ET ships are now being used to generate this level of energy.

"A lot of technology is involved in some kind of energy transference from the Grey world to Earth for the use by subspace humans. The energy is actually being transferred along some kind of conduit or channel to the human subspace realm. There is active cooperation between the two groups. The Greys have the capacity to produce the level of energy required to transform the entire planet at once in a short time.

"The energy is being used literally to bathe all of Earth with a subspace glow. In a sense, the planet is being irradiated. Enveloping the planet with intense subspace radiation acts to enhance the intuitions of the physical humans so that they can be more receptive to information coming from their own subspace aspects.

"The idea is that physical humans have a deficiency in being able to recognize their subspace selves, and the energy is being used to amplify the very weak mind-body connections of humans.

"Hold on, I am getting a strange piece of information now. I am perceiving that somehow humans have gotten perverted because of this weakness. This sounds weird, but I am getting some kind of overlay that there was a subspace disturbance of some kind, and the human evolutionary track got off course. I am getting something like the Lucifer Incident or rebellion as an AOL from the signal line."

MONITOR: "Move back to the humans in the structure."

CB: "These humans are real power brokers. They have been totally contaminated. They are dwelling in a false reality of self-gratification. They are causing lots of problems for others, it seems. The subspace beings are concerned about these particular beings' future, but that is very secondary now. They [the physical humans in the structure] can kill themselves if they want, but they must not be allowed to further damage others.

"I have to go back to an idea; it seems important. This entire thing is a residual from something that comes across as the Lucifer

Rebellion. Don't ask me to explain this; it just feels clearly like it. These humans are very similar to Greys before their early civilization decline, when that world fell due to the chaos in the subspace realm."

MONITOR: "Courtney, cue on the influence of the subspace beings on physical humans in the structure."

CB: "This is very operational. The emphasis is to obtain behavioral change with regard to a few critical decisions that they are making. The subspace beings are injecting thoughts into the wily, crafty, and nearly evil minds of these physical humans.

"The ideas need only be cemented for a short period of time during this meeting. Selfishness will again triumph for these folks, but at least some helpful and crucial decisions will have been made without these folks knowing why they did it.

"These subspace beings do a variety of tasks. The idea of bathing the planet in subspace light is supportive of their activity as well. The joint project is to positively influence as many people as possible by enhancing the subspace environment while simultaneously preventing some evil humans from making decisions that could enslave, capture, destroy the many."

MONITOR: "OK, Courtney. Let's end the session here. The target was 'Star Trek/idea genesis.' "

DISCUSSION

Immediately following the session, I made a summary interpretation of these data. I did this at that time because the memory of the session was still fresh in my mind. In these summary and interpretive comments, I noted that the subspace light that originates from the Grey world and bathes Earth acts to make the populace (and thus the audience) more receptive to ideas like those contained in *Star Trek*. The power brokers are those who decide which shows to fund. My monitor noted after the session that entertainment decision makers often meet at retreats and resorts to make such decisions. Without the promise of profit, no show will get production support. The power brokers have no reason to support a show like *Star Trek* unless it performs well and makes a profit.

The session described above must be interpreted strictly in terms of the specific target cue. The cue emphasized the genesis of the idea of the series, which, in retrospect, means the entire series rather than the information contained in one or more episodes. It seems to me that ETs generated the idea of *Star Trek* to transform humankind in some way. They simultaneously put their ideas in an appealing medium, ensuring its wide transmission, and readied human viewers with a tendency to receive that transmission in a positive way.

Some readers may object to my analysis, claiming that the *Star Trek* series was not watched by the entire planet, and thus could not have a significant transformative influence on the development of human culture. In my view, it would be incorrect to assume that *Star Trek* is the only vector of influence the ET or other subspace beings would use to redirect our culture. It is best to think of the results of this session as an analysis of only one of perhaps many vectors through which our culture is swayed.

My session left unclear the matter of whether and exactly how the ETs directly influence the specific content of television shows such as *Star Trek* on an episode-by-episode basis. I decided to refine the cue and conduct an additional session, solo. Thus, the cue that I used was "*Star Trek: The Next Generation*/episodes' idea genesis."

Date: 11 September 1994
Place: Atlanta, Georgia
Data: Type 1
Target coordinates: 3850/3054

Following the preliminaries of SRV, I began to detect colors like browns and tans. The textures were of wood and cement. The temperature was warm, and I smelled something pungent. My location gave me the impression of being flat and expansive. My Stage 3 sketch resembled a spread-out city.

In Stage 4, I perceived a congested city in a desert environment, and I had the strong AOL of Los Angeles, and particularly the Hollywood area. I executed a movement exercise to place me three feet beside the target (however it would be specifically identified by my unconscious), and I found myself in a bedroom near a sleeping human white male. I made a sketch of the person.

Entering the man's mind, I found that he was dreaming. I detected an implanted object in his brain and began to investigate that more closely. It was necessary for me to give myself some flexibility in time so that I could determine how the object came to be in this person's brain. Moving backward in time, I was able to determine that the object was placed in the man's brain by a Grey using a long surgical needle during a UFO abduction.

Returning to the sleeping male, I re-entered his mind and observed its dream state. It was forced, in the sense that he was being guided by information coming from the implanted object in his brain. This device monitored and inserted ideas on a regular basis, with a great deal of activity occurring during sleep. The person was unaware of the origin of these ideas, or of the presence of the device in his head.

Upon waking, the person regularly felt excitement with regard to his new ideas. He credited his own creativity for these ideas.

I then executed another movement exercise that placed me a thousand feet from the point of origin for the transmissions that were being made to the man's mind. I found myself near a bright circular light. At first I thought this light was an ET ship, since it had the clear signal line sense of being an advanced extraterrestrial device. But upon closer inspection, I found it to be a rather small (relative to an ET ship) mechanical device that did not appear to be visible in the normal physical sense. The device seemed to have a matter density just below that which can be seen by human eyes. It had the ability to move quickly, and it had access to an impressive amount of energy.

Moving my mind into the device, I clearly detected a sense of consciousness. There were no beings in the device, but the device acted as a conduit or portal for the transmission of consciousness (or perhaps just thoughts), a relay mechanism.

Moving into Stage 6, I traced the transmissions from the circular brightly luminous device back to the dreaming mind of the sleeping man. After retracing the path back to the circular luminous device, I then tracked the flow of the transmissions to the ultimate point of origin.

At this stage, I found myself in a structure on a planet. There were beings in the structure, both Greys and other humanoids, some looking quite like Earth humans. While the Greys looked solid, most of these other beings were of the luminous variety, which meant that they were generally subspace beings. The non-Grey humanoids wore white robes, and I got the impression that this was a Federation operation. I was able to determine that this operation was a joint project directly involving many species.

Probing further, I determined that the site itself was formally connected with the Federation authorities. Indeed, the beings involved with this site reported to Federation headquarters directly. I got the distinct impression that I was actually in the Federation headquarters, but in a different room than that to which I had gone in the past.

I then cued on the content of the transmission itself, and I determined that the message contained a tremendous amount of detail. The transmission included plot ideas, characters, pictures of specific scenes, images of planets, ships, and beings. The content was data that would later find its way into a human-written script of a specific *Star Trek: The Next Generation* episode. Not that the show was to portray ET beings exactly as they actually exist—it simply was to get people used to a variety of physical forms and cultures.

DISCUSSION

It is important for me to emphasize that I do not have any idea as to the identity of the Hollywood person whom I remote-viewed and who had the implant. I do not know if this person was a screenwriter or someone else associated with the creative process of a particular episode of *Star Trek: The Next Generation*. The person could even have been a friend or the spouse of a person more directly involved with plot construction. All that would be necessary is for this person to be in a position to formally or informally suggest plot and other ideas to people who create the series. Moreover, I do not know to which episode this remote-viewing session refers. It could refer to one or a number of episodes. I did not explore these questions further.

ETs are definitely involved in shaping planetary public opinion in a way that will facilitate eventual open human recognition of—and interaction with—extraterrestrial life. I strongly suspect that the *Star Trek* series is only one of many human ideas or events in which ETs have been involved. They want to help humans gradually stop thinking that we are alone at the center of the universe, and to understand instead that we are but one group in a complex galactic society.

Readers should be aware that I did not discern any attempt at crude mind control associated with this project. Rather, the ETs were listening to the ideas originating from the person's mind, and, during sleep, feeding him new plot ideas relevant to his job. The person was not forced in a malevolent way to accept these ideas. Instead, he had the free will to accept or reject these ideas but freely chose to accept them because of their interesting audience appeal. Indeed, he was elated to have woken up with such good ideas for a script. He had no idea that the ideas were not really his.

How far this ET project extends into human culture, I do not know. Personally, and on a totally speculative basis, I would not be surprised if others variously involved with television and movie science fiction production have been unwittingly involved in the project, either personally or through their associates. The investigation of this would be a good project for the next generation of scholar–remote viewers.

CHAPTER 18

A Return to Jesus

At this point in my research, the Greys confused me. I could understand that their genetics program could serve a variety of real needs. But all of my remote-viewing research efforts pointed to the idea that obtaining a different electrical/chemical/mechanical form was associated with something greater than having a good time. Indeed, when I remote-viewed the Grey mind, the issue of their own evolution was tainted with the sense of panic, as if the matter was one of life and death.

I now wanted some straight talk as to what the Greys were up to. I needed interpretive help with some of the results of my most recent sessions. I wanted it from someone who I hoped would give me thoughtful advice. Are the Greys after a better ride on the roller coaster of physical life, or are they seeking something more important? If there is something more important, what is it? Why should humans care? To prepare for this session, I developed a list of questions that would help me understand the importance of developing cooperative relations between humans and Greys. Basically, I wanted to know why humans should help the Greys. Put crudely, what's in it for us?

To get answers to these questions, I decided to return to Jesus one more time, and I decided to do the session solo, under Type 1 conditions, since I already had had signal contact with the target. The last time I targeted Jesus, I simply did not have the same questions that I now needed answered. I felt I was at a crucial point in my research. I needed guidance so that I could place what I had learned recently about the Greys into a broader picture.

Readers should be aware that it was necessary for me to formulate my questions in advance of this session. The reason is that it is not possible during an SRV session to engage the conscious mind to the extent that would be required to formulate questions on the spot without compromising the quality of the data. Thus, I began this session in search of answers to a list of questions. But I had no expectations regarding the outcome of the session. In true SRV fashion, I would accept the data however they arrived, saving analysis for afterward.

Date: 11 July 1994
Place: Atlanta, Georgia
Data: Type 1
Target coordinates: 8863/8473

This time the preliminaries indicated that my approach to Jesus was in a setting in the distant past. By Stage 4, I found him as a luminous being, but he seemed surrounded by physical humans. As it turned out, he was either conducting or observing a meeting at the time. He left the meeting, temporarily moved away from the humans, and stared straight at my subspace face. I got the clear sense that he was aware of and ready for my questions, and he wanted me to proceed.

I began by asking Jesus again if he wanted us [humans] to work with the Greys regarding their genetics project. His response was nearly a command. He categorically stated that we *must* work with them. They are God's children, no less valuable than those we call human.

I asked him if the Grey project had something to do with a greater evolutionary goal, like merging with God in some way. He responded in the affirmative. He then told me that their program is to enable their physical-mind circuits to recapture the flexibility required for individualized personality development. That development is what is needed for God consciousness. But he wanted me not to misinterpret him. In their own way, they are already close to both God and himself. They are loved.

Sensing more behind his response, I asked if the attainment of a fully developed individualized personality is a requirement of Grey evolution toward God. To this he replied both yes and no. God loves them, and will care for them. They made a choice to evolve toward God. God will not let them perish. But they have chosen the route of individualized personalities. They have been impressed with the lives of individuals with whom they have interacted, and they desire this road toward completion.

It was then necessary for me to understand what exactly is meant by merging with God. I asked him what it felt like to merge with God. He told me that there is no instantaneous change in personality. The primary difference is that a person's degree of perception is expanded.

I asked if fusion with God is the same as achieving God consciousness, or an experiential awareness of God as a sentient life force in all things. He then said that God consciousness is one and the same, however it is achieved. One is either with God in perception or one is not. One cannot perceive God without being fused with God.

I asked him if meditation can lead to this fusion with God. He replied that this is the goal, but that meditation is not the only, or even the primary, way that this is done. Normally, the route takes many life experiences and a great deal of time. Meditation is valuable only in the sense that it can shorten the process. This is true of humans as well as nonhumans.

Returning to the subject of the Greys, I asked if they had merged fully with God. He said not yet. They lack a sufficient degree of developed personality in order to fully share the God experience. They must proceed out of their situation in order to fully merge with God.

I then told Jesus that I did not understand the idea behind Christianity. Did non-Christians need to call on Jesus in order to evolve fully? Jesus' response to this question contained a sense of exasperation, and this is the only time I have ever experienced him so upset. Quite forcefully, he stated that a name is nothing. Everything depends on personality development, and this includes the development of a deep ability to both perceive and love beyond the self. In this respect, the Sidhis are valuable, but they are not Christian in origin. Understanding and loving God is important. This is what carries one through evolution.

(At this point in the narrative, I feel it is important to insert that Jesus did not tell me exactly how one goes about loving God, or even who or what God is. But I got the clear impression from Jesus that creating a mood about how wonderful God is had nothing at all to do with what he was talking about. Unfortunately, I did not think of asking him about these things before the session began, and thus I was not capable of thinking of these questions during the session due to the structural limitations of SRV.)

I then asked if God wanted to merge with the Greys, and Jesus stated that it was the Greys who desired to merge with God. It was their greatest act of free will. God offers the possibility of union, but the Greys freely chose to do this. He then emphatically stated that the destiny of the Greys is clear: they *will* merge with God.[6]

---

6. My use of words such as "fusion" and "merge" with respect to God is not strictly ac-curate, and their use is really a function of limitations in language more than anything else. When I refer to these things, I am really referring to an ability of beings to per-ceive and productively interact with *all* levels of life, even those beyond the realm of physical and subspace existence.

CHAPTER 19

Not All Greys Are Equal

The following session was unplanned, in the sense that the target was not on the list of targets for this research. My monitor gave me this target under Type 4, monitored and blind, conditions because his own intuitions suggested that we needed to know more about the abduction experience. Interestingly, after it became clear that we could now remote-view UFO abductions, the interest in this type of target greatly diminished for both of us. It is probably due to this that my monitor felt the need to return to the purpose behind UFO abductions before we totally redirected our attention elsewhere. My own unconscious approached this target from an angle that revealed to us a new aspect of the community of Greys that we previously knew nothing about. Thus, this session did more than incrementally add to our understanding of the abduction phenomenon. It helped us to more completely comprehend some of the complexities of Grey society, and it helped us to discern why there seems to be such a wide range of experiences associated with abductions.

Date: 13 July 1994
Place: Atlanta, Georgia

Data: Type 4, remotely monitored
Target coordinates: 7646/1231

The preliminaries of the session clearly indicated that I was on a planet that held a large saltwater ocean. My initial location was on the surface of the water. No land was in sight. Following the session's preliminary procedures, I detected beings working under the surface of the water. There, I encountered an underwater structure.

CB: "This is some type of hollow structure down here. It is metallic, and it appears to be a chamber of some sort. There are pipes in the structure, and a floor for walking on. This place gives me the creeps. It is really weird. I better put that down as an aesthetic impression [AI]. The entire thing is like a submarine. In fact, this is a pretty strong AOL from the signal line.

"Getting more detail on the structure. . . . The surfaces inside are predominantly metal, but there is some leathery material as well.

"There are four beings inside the structure. They look human. I am inspecting their clothes now. The clothes are of the human work variety, sleeveless T-shirts, normal slacks, and so on. They are definitely hard at work; they are sweating."

MONITOR: "Cue on their work activity. What are they doing?"

CB: "Hold on. . . . Fish. Something to do with fish. These are all male workers. They actually have no real understanding of what they are doing or what they are involved in. This is very complicated. They are doing things robotically. They are even unaware of their total environment."

The monitor has me move to Stage 6, where I explore a timeline that contains the individuals in the underwater structure. Earlier on the line, I find the individuals with normal mind states. I then go further back in time to locate them before they were born in physical bodies. I then enter the minds of the individuals in the structure to obtain a clearer idea of their mental condition.

"The fish are around the outside of the structure and are the focus of attention with regard to the humans. But there is a false relationship involved with this scene. The humans think they are after fish. But the real purpose of their activities is entirely different. There is no direct relationship at all between the actual fish and the humans.

"These humans may be captive, at least temporarily. It is like their physical bodies are under mental anesthesia, in a sense that their own thoughts are not controlling their bodies. They are not fully in control of themselves or their surroundings.

"Their minds seem to be on hold, as if working in a trance. They are working quite frantically, but they are not aware of this in the normal sense of human awareness. I am not getting a good feeling about how these folks are being used, and perhaps abused. It is almost slavelike, but not totally. Perhaps guinea pigs would be a better analogy.

"OK. I am expanding my view now. I am picking up on an ET presence. There is a *very* large ET ship involved with these humans. The ETs are Greys. The ship is very modern with a lot of technological gizmos. Actually, the technology is a bit odd, in the sense that I think it is not so advanced that humans would have difficulty understanding the hardware. I have not always gotten this sense from Grey ships. Anyway, the control of the sailors is from this ship."

MONITOR: "Courtney, cue on the first Grey-human contact for these people."

CB: "Doing that now. . . . There was an abduction experience for at least one of these individuals when he was very young. It could be the same with the others, but I am focusing only on one now.

"It seems that the Greys were involved with this individual in the prenatal stage. Following the signal, I found the human mother. I am now in what appears to be an Earth scene near a coast. The Greys put the fetus in the womb. Hmmm. These are relatively primitive Greys. They implanted the fetus physically in the womb. They did not use any high-tech means of insertion. It was a rather crude gynecological procedure, like the type humans could do in the not too distant future.

"These Greys seem less evolved. They tend also to blunder with human interactions, hurting humans without wanting to specifically. They just do not understand. They lack human compassion in a way."

MONITOR: "Cue on the concept of DNA."

CB: "Hmmm. It seems like the genetic makeup of these individuals in the submarine is what is being experimented with. Nearly all of their DNA is human, but there is a fraction that is not human, perhaps Grey, in origin.

"Hold on. These particular Greys are now becoming aware of my activities. It is odd how long it took them to notice me.

"Continuing with the Greys. . . . These Greys are working with small changes in the genetic makeup of humans. The changes are not necessarily Grey of origin. The genes can come from elsewhere as well. But the changes are really small and select. This is the same genetics program that we have been witnessing all along, but it is being conducted at a more primitive level, perhaps their early attempts."

MONITOR: "Are you saying that there are different groups of Greys?"

CB: "They are not really like factions, but they are different. This group is the most primitive that I have found working with humans. They are using humans as guinea pigs to see what happens with these minor genetic changes. The changes are directly oriented around mind control and/or telepathic communication."

MONITOR: "Why are they doing this?"

CB: "To modify the human genotype such as to eventually, in the long run, help themselves produce a new Grey vehicle. Actually, the word *vehicle* makes more sense than the word *body*. It is as if the Greys are primarily interested in the enhancement or development of group consciousness [Grey style] with human genes, sort of like setting up a huge human telepathic party line. This seems to be their first priority."

MONITOR: "OK, let's end the session. The target is 'Greys-humans/natal/prenatal connection.' "

CB: "Huh? Where did that target come from?"

MONITOR: "It was one of my surprises to keep you on your toes."

DISCUSSION

Two fascinating elements of the abduction phenomenon have been suggested by the data in this session. The first is that some abductions are not as benign as others. I do not have any data that suggests malice on the part of the Greys. It is easy to confuse incompetence with malice; abductees may feel that their interactions with the Greys are simply bad, and to them, the distinction is irrelevant. It does seem that incompetence on the part of some Greys is the cause of the uncomfortable relationship some of them have with humans. They do not seek to do harm; it is just that at least some Greys do not know how to interact with emotionally complex beings such as humans. They probably do not see their activities in the negative light that we associate with captured guinea pigs used for experimentation. Indeed, given my perception of the relative lack of emotional flexibility on the part of the Greys involved with this session, it is entirely possible that they treat humans in a way comparable to the way they treat themselves. You will recall that the idea of individual self-determination does not seem to click with them.

The second fascinating element of the data in this session is that there are apparently different categories of Greys who interact with humans. Perhaps we might call these categories (1) primitive, (2) advanced, and (3) superevolved. (As of the date of this session, I had not yet observed any of the superevolved Greys. I do so in a latter session, however.) The Greys in this session seem to be of the primitive variety, but this type should not be confused with the very early Greys who existed on their original homeworld just before the collapse of their planetary civilization. Those very early, planet-bound Greys did not interact with humans at all, to my knowledge.

Do all three primary types of Greys interact with one another? How do they organize themselves? When dealing with beings who can send their spaceships through time with technological ease, it becomes mind-boggling to try to figure out how the same species interacts with itself when both the current, past, and future beings are in one place. I can only guess that technologically advanced ETs are used to such things and eventually establish working protocols of interaction.

I ended this session with a deep realization of how much we humans still have to learn about the complexities of galactic life. It is not just that we need to learn about other societies that are spread throughout the galaxy. We also need to understand how species and cultures interact across time.

CHAPTER 20

Adam and Eve

In an earlier chapter, I report on a monitored session in which the Midwayers were the target. During the years of military remote viewing, the idea of targeting the Midwayers was a bit of a lark. Someone was reading *The Urantia Book*, a book of supposed revelations regarding the organization of subspace life, and decided to conduct a test. In *The Urantia Book*, the Midwayers are a group of subspace beings who live and work on this planet. No one in the military knew whether to take this idea seriously. But quite to everyone's surprise, repeated attempts under Type 4 data conditions revealed the same information. These subspace beings really do exist. In fact, the discovery of such kinds of beings led to a number of credibility problems among the higher-up military staff. It was hard enough to convince the likes of generals and admirals that trained remote viewers could observe missiles in a silo, let alone that there were a group of invisible yet friendly folks who like helping people in their own evolution.

It is important for me to state that I am not endorsing *The Urantia Book* in any way. I do not know how much of it may be true. My own investigations into the matter suggest that much, but not all, of it is accurate. It seems to contain false and manipulative information interlaced with truthful information, and it is not an easy task to separate the two without extensive remote-viewing probes. For example, the book seems to go to great lengths to deny the possibility of past lives in a way that remote-viewing data (not presented here) finds clearly fraudulent.

Nonetheless, because *The Urantia Book*'s discussion of Adam and Eve has a distinct ET flavor, I decided to have my monitor add this famous pair to the target list. If the book was correct in what it said about Adam and Eve, then it would explain a great deal about long-term ET intervention in the evolutionary affairs of life on this planet, and the targeting would thereby be more than worth the effort. It was a risky long shot, in the sense that we only had time for a limited number of monitored sessions, and a wasted session on a fraudulent story would be dearly felt. But as it turned out, the story of Adam and Eve in *The Urantia Book* is basically accurate. I report here on two sessions in which the target was Adam and Eve. One session was conducted under Type 4 data settings, whereas the other was conducted under Type 1 conditions. The second session under Type 1 settings was conducted to answer a few questions raised by the earlier data.

Some readers may wonder why the religious figures of Adam and Eve would be a target of investigation in a book about extraterrestrial civilizations. The basic reason for my interest in this target rests with a hypothesis that many of our human myths may have some basis in history. It is not that the myths have a close parallel to the actual course of events. Rather, the myths can potentially contain meaning about people and events that early civilizations little understood. The investigation of these myths using remote viewing can sometimes unravel delicate connections between the actual past and the stories about this past that have been passed down through the ages. *The Urantia Book* reveals just such an occurrence with regard to Adam and Eve. It was my desire to find such a myth-reality parallel, perhaps helping to explain some of the questions that have been raised in much of the scholarly literature regarding sudden changes in the evolutionary path of humankind.

I am not assuming, nor is it important, that readers have any familiarity with the discussion of Adam and Eve that is found in *The Urantia Book*. I mention the book only to identify the source of my original query.

Date: 14 July 1994
Place: Atlanta, Georgia
Data: Type 4, remotely monitored
Target coordinates: 5328/6080

The preliminaries up through the sketch in Stage 3 indicated that the initial approach to the target revealed a forested area near a mountain. A fast-moving structure was above the mountain.

CB: "I am perceiving some structure that has associated with it explosive energetics. It is fast and circular. It is moving on a curved and somewhat erratic trajectory from high to low, moving to something that looks like a partly forested mountain. I am perceiving a fir tree. Also, I am getting the AOL from the signal line that this place may be near Santa Fe Baldy in New Mexico.

"The object itself is an artificial structure. It is hard, and highly refined. There are windows or view ports. I am now perceiving beings in the structure, pilots.

"These beings are not Greys. Hold on. They are not Martians either. They are humanlike, but they are not current humans."

MONITOR: "Cue on their gender."

CB: "They are both male and female. They are wearing uniforms. I am going to put this down as an AOL of the signal line that they are advanced human beings. They appear to be humans from the future."

MONITOR: "Cue on their purpose."

CB: "These folks are here for observational purposes. They make absolutely no human interventions or contacts. They report directly back to the Federation. Also, they are not aware that I am remote-viewing them."

MONITOR: "What else do you see regarding the ship?"

CB: "The ship is mostly filled with flying apparatus. The only medical equipment here is for their own use in the event of an emergency."

MONITOR: "Return to the occupants. Get information regarding who they are and how they live."

CB: "These humans are advanced, but not much more advanced than current humans. Probing. They apparently are vegetarians. They have filled up their ship with food at their base. Food comes from organic vegetative sources in space- and planet-based gardens and from warehouses on planets. To get food from Earth is problematic. The problems are disease, and the possibility of disrupting human activity."

The monitor has me move to Stage 6, where I locate the target time on a timeline. The current time of the session is very near this target time. I begin to probe the concept of time with regard to these beings.

"They do not have that much flexibility to move ships back and forth in time. They are not like the Greys in how they can fluently move in time. But these folks in the ship do make regular observational visits to Earth. But again, in comparison with the Greys, it is not constant like the Grey activity."

MONITOR: "Locate their first visit on the timeline."

CB: "Hold on. Wow! That is an AI! I am getting the strong AOL of the signal line of Adam and Eve. These beings have been dropping in and observing for a long time.

"It seems that these folks formerly operated actively on Earth with a genetics program. These are scientists and technicians. They are now observing to see how work proceeds, but they are not allowed to intervene. They have been caught up in an Earth project for a long while."

MONITOR: "Cue on their initial Earth contact."

CB: "Initially, these folks were very naive, fresh out of training. They had some experience, but not a tremendous amount. Actually, their previous human bodies do not differ much from those that they have in current time. There seems to have been very little evolution for them. I am really picking up the idea that they are scientific managers or technicians of some type.

"Also, my mind is focusing on a particular couple. I am getting the overwhelming AOL of the signal line of Adam and Eve. I do not know if I can continue the session much longer through this AOL given its strength."

MONITOR: "It's OK. We can end the session. That is the target, 'Adam and Eve.' "

A few months later, I still felt that I needed more information regarding the original Earth activities of Adam and Eve. Bluntly, who were they and what were they doing? Thus, I conducted a solo session under Type 1 conditions in which the target cue was "Adam and Eve/original Earth activities."

Date: 16 September 1994
Place: Atlanta, Georgia
Data: Type 1
Target coordinates: 6957/4096

The preliminaries indicated that the target involved a considerable movement through time, solid land, and some man-made structures. The temperature was quite warm. The smell was both human and organic. I heard the sound of voices.

Following the preliminaries, I noticed that the climate was pleasant and basically dry. The environment seemed Middle Eastern, Mediterranean. I noticed two types of people, light- and dark-skinned.

Soon afterward, I got the sense that there was a power source of tremendous energy near the target site. I had the AOL of the signal line of a nuclear reactor of some type.

Together with the idea of concentrated energy, I had the sense that some of the beings near this site may not be happy. I received the AOL of the signal line of a slave camp, and the sense of repression. I did not pursue this, nor did this AOL return during the remainder of the session.

Around the site were small-scale machinery, stones, and buildings. Most of the inhabitants seemed to be living a calm life. There was no sense of crisis. Things were peaceful, but there was an underlying tension in the air.

I cued on the idea of tension and located a luminous nonphysical being related to the site. The being seemed like a military leader of some sort. There were other subspace beings near and around this being.

Associated with this leader was the idea that a division had recently broken out. I had the AOL of the signal line of a war. There had been a heated argument and many people took sides.

This planet was out in the boondocks, far from civilization as these beings knew it. One side felt that they should look out for themselves and disregard the distant authority. Two camps formed. The minority camp was loyal to the distant authorities, and they exhibited a significant degree of bravery in their interactions with the other side of the dispute. There was a long standoff as the two sides separated their contacts with one another.

I then executed a movement exercise that placed me three feet from the specific target. I found myself on what appeared to be a beach near the sea. A male and female were present. Farther away, there were a number of other advanced and seemingly pleasant beings.

I moved my mind into the deep mind of the male. He felt isolated in some sense. He was alone, and in love, but not in a puppy love sense. He was mature. Shifting to the deep mind of the female, I perceived that she was a manager of some sort. She was a very advanced worker, and she was also very devoted to her husband. However, there was a strong feeling that she needed to push her husband to enhance his career, in the sense that she felt he was not advancing fast enough.

I then cued on the situational problem at the site. There was a subspace disturbance. There were beings at the site who were rather desperate and who needed to disrupt the smoothness of the operation. When I probed the idea of leadership, I got the sense of military structure, and the AOL of the signal line of the Lucifer Rebellion. I did not pursue this idea. But I did get the sense that this place is generally distant from the center of authority, and that this rebellion was primarily opportunistic.

Shifting my focus to the work of the male and female at the beach, I discerned that they both taught and organized. Following the signal, I found that they worked with the local humanoid types. They taught subjects related to reproductive behavior and mating preferences. They also taught skills, mainly to generate interest in reproduction-related lessons. Their intention as teachers was good, in the sense that there was no malice. But the purpose had an unsettling sense of selective breeding that had both passive and active components.

I cued on the purpose of their activities, and I discerned that they wanted to engineer a new and unique race. They could not simply put their own genes into the pool. Rather, they planned to speed up nature without forcing an overcontrol of evolution, to get the end result of many generations of natural mutations and sorting and weeding in one fell swoop.

The problem was that the program went astray. The project was not well planned from the start. It was a shoestring effort with little or no supervision. Too much reliance was placed on individual loyalty and common sense. These folks were placed out in the boondocks, and they got intellectually disoriented.

Cuing on the program goal, I received the clear sense that these beings were playing God. They were in a rush to change things along lines that they approved. But in their deep minds there was also fear.

They feared that God, left alone, might come up with an evolutionary being greater than themselves. There was an element of pomposity in their project design. Somehow, the tension that erupted out of the project was due to a flaw in the original plan of the society that originated it. The motivation behind the idea of rushing evolution was a mistake. This caused a crisis in the collective consciousness of these beings, such that the stress to rush and produce advanced sentient beings became addictive to some elements of their population. These other elements became sick and perverted, missing the original purpose behind evolution as a process leading to something greater than themselves. In one sense, the sick elements of the society saw themselves as good, and wanted to convince both themselves and others of their worth by pushing the evolution of the other races in *their* direction. They saw themselves as evolutionary end-products.

Adam and Eve apparently sided with the minority of their comrades who were loyal to the more distant authority. The rebellion was short-lived.

DISCUSSION

Adam and Eve were project managers in a genetic-uplift program for humans on Earth. They were not naked simpletons who ran around in the woods. The myths that surround these two beings have undoubtedly been inspired by the intuitions of seers who had some natural remote-viewing abilities, but who could not place the activities of this couple within their own intellectual confines. Thus, the couple became known as the beginners of the human race. In a sense, I suppose this is partially true, since they were involved in a project to manipulate the human gene pool.

Adam and Eve are still alive today, in both a physical and subspace sense. They do not, supposedly because of Federation rules, interfere with the activities of the Greys. But they have a keen interest in the outcome of the current genetic trajectory of Earth humans. Obviously their bodies are not likely to be the same as those of long ago. It is interesting, however, that I did not detect a significant evolutionary advance in their physical forms. I got the distinct sense that something was not right with this couple, although I am not sure they would agree with this.

Nonetheless, one thing is certain. The genetic manipulation of the human species is not a new idea. It has been going on for a long time. Moreover, this may be one of the primary reasons behind the phenomenon that some evolutionary biologists call punctuated evolution, in which an evolutionary trajectory takes on a new direction relatively suddenly. Much more research (using both remote viewing and traditional scientific methods) needs to be done to verify this tentative hypothesis.

CHAPTER 21

Guru Dev

Early in our research, both my monitor and I were becoming convinced that there was much more to this project than the simple investigation of who was flying the saucers. By the summer of 1994, we had obtained remote-viewing corroboration of the abduction phenomenon, and we were becoming fairly well versed in the ideas underlying the basic genetics program of the Greys and the problems that are being faced by the Martians. However, sensing a much bigger picture, we agreed after much discussion to solidify our earlier but tentative decision to include other targets of wise beings besides Jesus that could give us advice as to how to interpret some of our data. This chapter is the result of targeting one such individual, and I conducted the following session solo in a Type 1 setting.

Guru Dev was the meditation teacher of Maharishi Mahesh Yogi. During the many months of my remote-viewing research, I was sensing clearly that I needed to ask Guru Dev some questions. Other remote viewers had been observing a group of Martians that they called the "priesthood." These Martians seemed to have some out-of-body travel and communication capabilities, and my monitor thought that maybe they did the Sidhis. The Martian priesthood was on our long list of targets, and I knew that I would eventually be given the target blind. But I wanted to get some information about them before diving my mind into their midst. If they did the Sidhis, I needed to know this, and soon. Thus, one morning in the summer of 1994 in Ann Arbor, Michigan, I targeted Guru Dev. Since it is a solo session, I report it as a narrative, thereby omitting nearly all of the jargon of SRV's protocols.

Date: 24 July 1994
Place: Ann Arbor, Michigan
Data: Type 1
Target coordinates: 3745/4021

The preliminaries indicated energetics, land, and something man-made.

My initial perceptions included colors such as blue, white, and brown. The textures that I perceived were airy. Again, and as with all SRV sessions regardless of data type, I had no idea how I would get to Guru Dev, or in what setting I would find him. The protocols of SRV are set up to force the unconscious to make all of these decisions. My conscious mind was just along for the ride.

The temperature was comfortable. I began to discern a sweet taste, and the sounds of a form of Indian music called Gandarva. In the subspace "air" there was the delicate smell of incense. I began to chuckle to myself: it seemed that Guru Dev was setting a stage.

As I proceeded with the protocols, I found myself in a place that seemed more subspace than physical. The topography seemed irregularly shaped, with dips and holes, like tide pools along the beaches of East Africa. But there was no water. I noticed that there was a sky overhead.

Slightly off-center of my view, a light being looked at me. I perceived this being to be the target and approached. I sensed that it was indeed Guru Dev, and he was waiting for me.

Before engaging in a conversation with Guru Dev, I looked around. I made careful observations of the surrounding environment. It was quite colorful, and I found myself remarking that the place was a bit weird (for me). The overall ambience was very comfortable, but I had never imagined a place that seemed so physical and subspace at the same time. It certainly was a place of special significance, although to this day I do not know where it was.

Redirecting my attention to Guru Dev, I noticed that he was wearing wraparound white clothing, although the color was not totally white. Indeed, the clothing had many shades of luminous colors to it. I telepathically told him that I had questions. He seemed to know this, and he indicated I could proceed.

Remaining within the confines of the SRV protocols, I asked him if the Martians have a priesthood. The response was clear: they do. I then asked if the priesthood does the Sidhis. Quite clearly, they do not. I immediately asked him what they worship. Interestingly, he indicated to me that I should find out from them. He thought I should experience it directly.

I then asked Guru Dev if the Federation council members do the Sidhis. I sensed that he became a bit more serious, and he informed me that they did something related, but not exactly the Sidhis. They did something that was appropriate for their own level and experience.

Continuing, I asked Guru Dev if the Sidhis are useful for a diplomacy course for human representatives to the Federation council. Following this question, I received the strongest response of all. Guru Dev emphatically indicated yes. Indeed, practice of the Sidhis will greatly assist humans with their interactions with the council members in the Federation. Guru Dev's emphasis related to the types of personalities that are on the council. Human representatives familiar with matters of consciousness would more easily interact with these personalities regarding matters of state than could representatives not so trained. I got the warning sense that humans should not mess around with the Federation by sending just anyone to the headquarters. It would be like the United States sending an untrained person to be the U.S. ambassador in Moscow. No one would take the person seriously, and Russians would eventually wonder what kind of people the Americans are. Humans need to send representatives to the Federation chambers who are actively engaged in their own accelerated growth in consciousness. Mature and rapidly evolving human representatives would speak well for their fellow global citizens on Earth.

I then asked if Guru Dev could see any problems with the use of scientific remote viewing as the means for representatives to communicate with the Federation council. He said that this method of communication is not optimal, and it will change later as human society matures. But for now, *it is the only way possible.*

At this point, I had run out of questions. I just looked at him and asked if he had any further thoughts. He just looked back at me, and indeed, into me. He was just calm, very, very calm. I thanked him and ended the session.

DISCUSSION

Following this session, I felt certain that I would be able to outline a comprehensive course of study for diplomats who would represent human interests in the Federation. Humans are not yet full members of the Federation, but representation could soon begin on an official level with fully trained diplomats. The galactic diplomacy course that I have constructed is outlined in a later chapter.

CHAPTER 22

God

To my readers, I have a confession: for this chapter, neither my monitor nor I could restrain ourselves. For a long while we tried to steer this book entirely away from religious topics. But the idea of evolution toward some central point reappeared wherever we looked. Moreover, religious themes kept overlapping with what we thought were simple ET concepts.

Prior to this, my monitor hadn't the courage to target God with SRV. But finally, he became much more daring and gave me a set of coordinate numbers for a blind, monitored session in which the target was God.

But basically, what we learned is that the idea of God is so great (in the sense of broad) that the unconscious can only convey direct information about God to our conscious understanding by way of metaphor and example. Apparently, God is no man on a chair in a fancy palace. I encourage readers to be patient with my presentation of the results that we obtained. Other remote viewers will target God in the near future, and our knowledge of God will increase as a consequence of these efforts. But for now, this is what we have.

Date: 27 July 1994
Place: Ann Arbor, Michigan

> Data: Type 4, remotely monitored
> Target coordinates: 3590/6110

The preliminaries indicated that the initial approach to a complex target was associated with artificial structures and liquid.

CB: "I am getting browns, tans, blues, and white. There is something splashy, also rough and dirty, soft. Temperatures are warm, tastes . . . salty. I smell something burning, acrid, smokelike. I am hearing machinery." *I sketched a structure near a body of water.*

MONITOR: "Go to Stage Four." *(He said very little for the remainder of the session.)*

CB: "OK. I am perceiving a Grey—in fact, many of them. There is something unusual here. These Greys seem to be early Greys. The eyes are a bit smaller, skin on the face is a bit pockmarked. They have genitalia. They are working.

"Wow. I just hit an emotional impression. They are hopeless. There is tremendous fear and a sense of desperation. I am getting mixed emotions from them. They have the sense like 'the sky is falling.' I am getting the clear perception that this moment in their history they feel is the end of their world as they know it. There is definitely a catastrophe going on here.

"Gosh, I am feeling sorry for these folks. I am putting that down as my own aesthetic impression and moving on.

"Wow, all hell has broken loose here. There is a rapid decay of their planet's ecosystem and their own ability to cope. There is widespread fighting. Yep, there are wars going on here. They're not really like human wars, but they are wars nonetheless.

"I get the sense of social organization based on clans and family groups. There is biological warfare. I get the sense of emptiness.

"I have just had a sudden time shift. I went forward in time; I do not know why. Continuing.

"I am now on a barren world. There seems to be some water near, but it is dead. I am on the surface of the land. The Greys are gone from the surface of the planet. There are underground chambers here now. The Greys have moved underground. The fighting has stopped, and the warring sides have declared a truce.

"The Greys are spending a long period of time underground.

They are moving to enhance their technology at a frantic pace now. There is a desire to move off the planet eventually.

"Wow. These folks are just plain tired. They are exhausted on the level of civilization. They are examining their basic essential nature. This was when they started to modify their genetic structure.

"They have new ways of manipulating and generating both physical and subspace energy. They're also close to achieving subspace transport. They are looking for a possible dimensional escape from their situation and world, not just a physical shift to a neighboring place.

"These folks are now underground, and they have a tremendous urge to see light. There is no natural light down there, and so scientists are investigating 'inner light.' They discover a new universe in subspace, pure and pristine, with civilizations seemingly better, more balanced than their own. They think that by escaping to this other dimensional realm, they could leave their physical problems behind. They would not be affected by the earlier problems. They made the mistake of thinking that their basic problem was in the texture of their existence and that other dimensions will not have these problems. I get the idea that they want to run away to subspace heaven.

"OK. Something new. That same old and wise Grey that has popped into a number of my remote-viewing sessions is now observing me. But he is not actively participating in this remote-viewing session. He's just watching.

"I am now cuing on the result of the drive to escape. OK. There is initial euphoria. There is great spiritual progress, as they see it. There are many advances in evolution along these lines. But they run into a dead end. I get the similar sense of an athlete who exercises only one arm or leg. The other limb atrophies.

"These beings are now unhappy. They see in other evolved beings a happiness and completeness that they lack. They see no other choice but to begin down a long and dangerous (from their point of view) path toward their past. But they are committed *not* to return to their former state. They fear the past. They will not get caught in an evolutionary loop.

"I am getting the sense that the human genetic project is the new escape route of a very long exodus. I am getting the comparable sense of the Israelites in the wilderness for many years after their exodus from Egypt.

"I am now cuing on the concept of race/destiny." *A one-second pause.* "I just had a big time shift forward. Wow! What is this?

"Hold on. What beautiful creatures! Wow! I am getting my own aesthetic impressions all over the place. This is amazing. These future Greys look much like humans. But they are like no other human that I have ever seen.

"They are telepathic. But there is more. They have learned to love. That is the basic emotion that they emphasized in their genetic project. These beings have love overwhelming. Their electrochemical machinery now fills this need of theirs.

"I am getting the analytic overlay of Jesus. Not in the sense that Jesus is here, but they are like an entire planet of Jesuses. The main difference is that when I remote-viewed Jesus, he had a sense of command or authority, whereas these Greys just have love without that added element.

"There are males and females here. They have sexuality. They are very healthy and the females give birth themselves [i.e., no canister births]. These people are very evolved and spiritually united."

MONITOR: "OK, Courtney, you can end the session now. As an aside, that was one of the most beautiful sessions I ever heard you do. It had tremendous flow. I basically said nothing throughout. What did you think of it?"

CB: "To be honest, it was very interesting. But I do not have any idea of what kind of target could have done this. This did not seem like one of our original targets on the list. What was it?"

MONITOR: "Sewage treatment plant, Fort Meade, Maryland."

CB: "OK, what was it?"

MONITOR: "God."

CB: "What?"

MONITOR: "God. That was the target."

DISCUSSION

In my view, God may have many aspects, of which we can, in our own ways, know only parts at a time. Our ability to understand each aspect of God depends on our own level of evolution toward what some call "God consciousness."

It seems that God is sentient, and that he (pardon the gender bias of the language) literally exists in fragmented form in evolving life and everything else. It is as if God experiences joy in creating matter and life from his own substance, and then living life through the experiences of species everywhere. This session leaves open how God initially fragmented himself in the first moment of creation.

One of the characteristics of God seems to be intelligence, variously manifested. The intelligence can be conscious, as in the sense of thinking. But it can also be automatic from our point of view as observers, in the sense of the automatic nature of our immune systems, or the cosmic dance of the stars.

This intelligence emerges in fragmented form in the shape of living and thinking beings. Once this occurs, the living beings, upon attaining a threshold level of self-awareness, begin to yearn to reunite with their source.

The weird thing is that these living beings initially seem unaware that they are literally made up of the substance of the source. Thus, less-evolved beings do not seem to understand that it is as impossible to fear God as it is to fear oneself. Yet the evolution of living sentient beings is eventually defined in terms of the beings' ability to discover this relationship between the original God source and their own selves. Once this happens, the dominant theme of existence is love. One can love oneself, and all others, because one sees that all of everything is created from the same fabric. Somehow, love is the theme of God, the glue that keeps the universe together. But only highly evolved beings realize the full extent of this reality.

I do not claim to know why love is a glue of the universe. We tend to think of love as a mushy emotion. My remote viewing of highly evolved beings seems to suggest that the human concept of love is very primitive, but I really do not know of any other word to describe the flavor of what I sense. Whatever love is in these advanced beings, it is not mushy. It is matched with clear thinking and effective action. There is a smoothness in their lives that is enviable.

Since love is pleasant, I consider ourselves very lucky. With my limited understanding of existence, I can see no logical reason why God could not have some other emotion—even hate—as a universal constant, other than the fact that it might yield a self-destructive terminal existence. I am sure there is a good reason for the dominance of love. I just do not know it. But I have one hint.

It seems that God experiences existence through his creations. Since all of everything is constructed of God's substance, our feelings and experiences are his as well. In a crude sense, we are like cells in an infinitely large body, and the infinitely large body experiences all aspects of the existence of each and every cell. Love is the dominant theme of the universe because it is natural (whatever that means) for God to love himself, not in a bad way, but in a healthy, growing, expansive way.

Note that in my own session, the Greys became highly evolved beings dominated by a sense of universal love that was a bit overwhelming to me. But notice also that the Greys did not collapse back into a point source and cease to exist once they realized that they were part of a larger entity. Rather, they became more like God in his more intimate nature, and they remained manifested in the universe as separate God fragments.

From this, I deduce that God has no intention of destroying the variety inherent in his expanded self by pulling himself back into the original point source. Rather, it seems that God's plan is to continue to expand, with little purified God-units running around in the universe.

Moreover, the evolutionary process may never end. Why should it? If all of everything is the fragmented manifestation of God, why would God want to stop the expansion of his own existence? Doesn't it make more sense to say that he will want to grow in loving variety throughout all eternity?

I am just speculating. But based on all of my remote-viewing observations, I see nothing in all of the universe that would indicate that God is planning on closing up shop. The most advanced beings would figure it out eventually if God were planning on collapsing everything. This would lead to universal frustration and an attempt to fight God's plan. Indeed, it would be like a cancer that could lead to universal suicidal tendencies.

But I do not see evidence of this anywhere. Wherever I look, I see beings struggling to improve, to find the meaning of existence.

And always, I find advanced beings whose existence is flavored with the intuitive sense of love, not the panic that would be the inevitable consequence of a fear of demise.

I think God has a plan to keep us around for good. This is my interpretive guess based on my observations. I perceive that we ourselves are manifestations of God in a very real sense. Currently we may be primitive relative to some of the more advanced folks in the universe. But we are on the upward track, and our life-and-death struggles in physical existence seem to produce an urgent desire in ourselves to understand our essential nature, thereby turning our attention toward the point source, and then, eventually, back to our selves in a cosmic mirror.

Quite honestly, I do not understand much more than this about God. I suspect I will enjoy struggling to understand him for the foreseeable future. I truly hope that God's mysteries are infinite, though, because if I ever felt I knew everything about God, and thus about myself, for what purpose would I continue to exist? But these days I must remind myself that I really have no idea what surprises tomorrow will bring.

CHAPTER 23

The Martian Priesthood

Long ago, when military remote viewers first started tracing the trajectories of Martian ships from the Red Planet to Earth, they noticed a certain group of Martians who played a unique role in their society. They seemed like medicine men, or shamans. They were highly respected by many other Martians, and more ominously, they seemed to have an ability to detach their subspace aspects from their physical bodies in order to attend gatherings of others similar to themselves. Quite honestly, this spooked the U.S. military.

Eventually I viewed these mysterious beings who had become known collectively in remote-viewing circles as the Martian priesthood. (Guru Dev had already told me that they did not do the Sidhis, but beyond that, I knew nothing of them.) This session was monitored under normal Type 4 conditions.

Date: 27 July 1994
Place: Ann Arbor, Michigan
Data: Type 4, remotely monitored
Target coordinates: 8711/3454

The preliminaries suggested a target on dry land and artificial structures.

CB: "I am getting tans, reds, and browns for colors. It seems sandy here, rocky. It is cold, cold, cold. Seems like Mars. Let me AOL that for now.

"I am on the Stage Three sketch. I have some low hills, a modest depression in front and on my right, and some channel cutting diagonally up through the middle. Could be a riverbed or something. Should I check for water?"

MONITOR: "Just go to Stage Four."

CB: "OK. I am in the matrix now. There are lots of rocks here, red rocks. It is basically a flat area in terms of any steep mountains. There is no air here. Barren, no life, on the surface, that is. This is a very harsh environment. Actually, I am picking up some air now. It is very thin, and dry.

"I must add that this place is very beautiful, austere, but in a backpacker's heaven sort of way.

"I am picking up beings now. They are not Greys. This really seems like Mars. I am putting these beings down as Martians, signal line data. I get males and females. They are very thin, light, and they have some wispy hair. I am trying to pick up their location now. Hold on.

"They are in chambers. I am in one now. These places are not so modern. There are basic survival necessities. Actually, these are quite sparse living quarters. Checking, hold on. These chambers are underground. What should I do now?"

MONITOR: "Cue on the idea of governance."

CB: "Doing that now. OK. These folks have a primitive organizational structure. There is no large-scale participatory network, like American voting. There seems to be a clan hierarchy. The elders have respect and authority. Survival conditions do not allow freewheeling democratic exuberance. It is a hierarchical, authoritative structure. The levels of authority are achieved through experience, combined with small-scale voting among the older elite. I am sensing to cue on religion."

MONITOR: "Do it."

CB: "There is worship here. It is used to keep the society together, given the harsh realities of physical existence. The children need it, as do the mothers. The elder elite are not so persuaded by it, however. But they do it lip service to support others, and out of a hope that perhaps the tradition will help them. It helps everyone on some level.

"I am cuing on religious leadership now, as I sense a pull in that direction. They are priestlike. Monks. They are the top of the social totem pole. I am sensing that they use symbols, and magic. They have a primitive but real understanding that there is more to reality than physical stuff. I am getting the sense that they are somewhat similar in certain ways to the mullahs of Iran, especially in terms of their internal organization. They influence both government and society. But they seem more like shamans than very evolved spiritual beings.

"I am following the idea of their magic. Hold on. There are objects, like totems or fetishes. Yes, these are like West African fetishes. They also definitely have out-of-body experiences that seem to reinforce their sense of religious concepts. It seems as if one such priest is now aware of this remote viewing."

MONITOR: "Locate the religious leadership."

CB: "They are all over, but in pockets. They are like a security apparatus as well. They spy using their own intelligence processes to maintain their control over the remaining masses.

"This is curious. Let me find out the limits of control for this religious leadership. Hold on. . . . There are two strata for this society. The religious leaders control the lower level. The bureaucratic-technical hierarchy tolerates them because they currently have no substitute belief structure for the masses.

"With regard to their location, they are on Earth as well. Hold on. . . . This is interesting. They see the move to Earth as an important power struggle. The struggle is during the shift from Mars to Earth, in the sense that there is not much trouble for them on Mars right now. On Earth the masses may abandon the religious leaders entirely.

"This is an authentic and very real struggle to maintain their Martian traditions. They fear not just loss of personal control for themselves, but the destruction of all tradition that makes their people who they are, i.e., Martians, or at least different from normal humans."

MONITOR: "Find out more about their out-of-body state, for a typical priest."

CB: "OK, give me a second. . . . They are summoned. Not a lot of business is transacted. They have the same difficulties that humans have in working between the two levels, physical and subspace. Subspace is used to communicate, though the priests communicate physically as well. The out-of-body state is for the bonding of the priesthood unit, and to keep the populace in awe of tradition. It also helps strengthen the sense of differentness with regard to human traditions. They seem to preach that this state is the advanced ability of their species and kind. They know that this is not true, but it is a useful thing to preach to control the masses."

MONITOR: "What about symbols of the leadership?"

CB: "I am getting that now. I will make a sketch. . . . That is funny, it seems to be the same symbol that was on the uniform of the Grey folks who came to rescue the Martians a long time ago. I wonder how they got hold of that?"

MONITOR: "Cue on meetings between humans and Martians."

CB: "The Martian priesthood is rather primitive. They do not like us humans, it seems. They want to segregate their populace on this planet. Hmmm. Looks like they are upset with the course of events, which seems to be moving away from them.

"OK, I am getting something now. The meeting will be soon, and with the bureaucratic-technical hierarchy, *not* the priesthood. I get the clear impression that the bureaucratic-technical hierarchy does not want us to deal with the priests. It is not the way for success in human and Martian interaction. For humans to contact the priesthood would be similar to the Martians contacting the Pope instead of the United Nations."

MONITOR: "OK, Courtney, good work. End the session."

CB: "And the target was?"

MONITOR: " 'Martian priesthood.' "

CB: "Figures."

DISCUSSION

This session covered a great deal of ground and answered a number of important questions that remote viewers had had for some time. First, the Martian priesthood is not the official governmental apparatus of Martian society. I do not now know the extent to which the competing authorities of the priesthood and the secular bureaucrats divide the society. Intuitively, I sense that the dominance of influence among the masses leans in the direction of the secular leaders, but that these leaders still cannot ignore the priesthood.

At this point in my research, I have only this minimal sketch of the dual vertical flow of the organization of Martian authority. However, it seems clear that there are two distinct levels of the masses, and that the priesthood's dominant share of influence rests with the lower stratum. What defines the two strata is not clear to me. It does not seem as if wealth is a criterion, but some other factor. I could guess educational level at this point, but I suspect that I may be in error, since I do not see any reason why Martians would deny equal levels of education to all of the members of their society, given their need to develop all of their populace in preparation for their eventual (or at least theoretical) migration here.

CHAPTER 24

The Roswell Incident

Media reports have long suggested that there may have been a flying saucer crash near Roswell, New Mexico, in 1947, in which live aliens may have been captured by the U.S. military. Some have claimed that at least one alien was held for a lengthy time as a prisoner on a military base before dying of unknown causes, and that at least one ship was carted off to military laboratories for analysis.

While in the military, my monitor once attempted to discern from military sources if the Roswell Incident actually did occur. The biggest problem was that no one could find either the dead bodies of the aliens or their spacecraft. Yet too many people in the New Mexico area, and some in the military, seemed sincere with their memories of such an incident. It puzzled him.

Eventually the military remote viewers were assigned the task of targeting the Roswell Incident. Initially, all did not go well. When the military team viewed the event, they saw light globes, not crafts, hovering low in the sky. They detected the "flavor" of ETs, though, indicating that ETs were involved in some way.

Following our breakthrough with abductions, my monitor and I decided to add the Roswell Incident to the long list of targets. The logic was that if the ETs had changed their minds in allowing us to see what was going on during an abduction, then they may have changed their minds with regard to other things as well.

Date: 28 July 1994
Place: Ann Arbor, Michigan
Data: Type 4, remotely monitored
Target coordinates: 7633/4128

The preliminaries indicated that the target involved flat, dry land and artificial structures.

CB: "I am getting mostly tans and browns. The textures are sandy, woody, and windy. The temperature seems warm. Wherever I am seems beautiful. I will sketch the place now in Stage Three." *My monitor gives me a movement exercise that places me five hundred feet above the target.*

"OK, it is sandy and rocky here, quite dry, warm to hot. I will draw another Stage Three sketch."

I sketch the picture of a structure in the air that is following an erratic trajectory, and another structure on the ground. I also sketch in some notable topographical features. My monitor gives me another movement exercise that moves me inside the structure that is in the air.

"There are black and grey colors here, glossy and clear, polished and smooth textures as well. I hear voices, but I cannot tell what is being said. The place feels very busy, however. It is very densely organized in a small area, very compact."

I draw another sketch of the inside of the structure before the monitor suggests that I move on to Stage 4 of the SRV protocols.

"Hmmm. There are some rooms here. There definitely are beings here. They are at work, and they appear to be frantic, not in a panicky sense, but almost. I am getting the strong analytic overlay that this is the Roswell Incident."

MONITOR: "Stay in structure. Put it in the matrix as an AOL and move on."

CB: "Well, there are four beings. These folks are now really terrified. Something has gone wrong. The machinery is broken. These are Greys, and I am getting a nearly overwhelming AOL that this is the Roswell Incident."

At this point, it seems that my unconscious is not going to let the session proceed without confirmation that the Roswell Incident indeed is the target. Since I am already thoroughly bilocated, there is little danger, and really no alternative but to acknowledge the obvious.

MONITOR: "That is the target. Stay in structure. Move quickly through Stage Four. Dump a sketch in the matrix of the inside of the ship."

CB: "Doing that now." *The speakerphones are silent for a moment while I sketch as quickly as I can.*

"The saucer is behaving erratically. They are losing control. They are experiencing fear. They know that they cannot be rescued. The problem is something about rules."

MONITOR: "Dump it in the matrix and move to Stage Six. In your Stage Six worksheet, draw a small, one-inch-diameter circle in the center of the page. Label this the mission origination point. Then draw an arrow to the destination point. Probe this and then dump what you have in the matrix."

CB: "Doing that now. Probing the origination point. This is a rocky place, craters. I am again getting a strong AOL-matching signal. This is the Moon. I am there now and looking up. Earth is in the sky. OK. I am now following the trajectory to the destination point. All I get is that the destination was a mission on Earth."

MONITOR: "Cue on the mission purpose."

CB: "Hold on. This is weird. It seems like the mission was to crash. The destination was Earth, and the purpose was to crash and thus force humans to investigate ET questions. I find this hard to believe."

MONITOR: "Don't analyze. Dump that in the aesthetic impression column: 'I find this hard to believe.' Move on."

CB: "The idea was to show that ETs are one, physical; two, vulnerable; three, not really different from humans; and four, can make mistakes. They knew the future, and thus knew they would crash. But knowing the future doesn't change the course of the event unless desired. The machine really did break, and the beings really did panic, crash, and die physically."

MONITOR: "Move forward in time and write down what you see."

CB: "Now I get humans on the ground. Military humans. These humans seem to be near panic. The military is running around, literally, and picking up every piece of the crashed ship. They're putting things in boxes and bags.

"There is the sense of ultimate urgency. Higher-ups are scrambling to contain the impact of the event, create secrecy. I get the sense that plans are immediately being developed to cover up the event. This reaches to the highest levels of the military. I am sensing that the president is being briefed verbally."

MONITOR: "In your Stage Six worksheet, draw a timeline with three points on it, points A, B, and C. Label point A the time when the entities were panicking, point B the crash point, and point C the time when there was no more debris on the ground. Then let's do a movement exercise." *The monitor then moved me to point C on the timeline, one thousand feet above the target.*

CB: "I smell burning flesh. I hear motors. There are vehicles driving erratically over the ground. Military vehicles. I am also detecting a single ET ship in the air. I am putting this all down in a Stage Three sketch.

"Moving into the ship. The occupants seem unable or unwilling to intervene. This ship is working smoothly. The occupants are in a panic about the activity below. I get the sense that they are watching as the 'barbarians' collect their colleagues."

DISCUSSION

In the remainder of the session, my monitor had me explore the idea that the ETs may have manipulated time with regard to the event. He explained to me after the session that he believed that the military had no physical evidence of the crash. He suggested that it may be possible that the ETs knew the future, and thus chose to let the accident occur without intervention. But then after the event had occurred, they went back in time to eliminate the event by preventing the crash.

I strongly objected to this hypothesis, since I had indeed remote-viewed the event. But my monitor continued to argue that there could be two timelines, one in which the event did occur, which is why I could remote-view it, and one in which it did not occur, which is why there is no physical evidence of the event.

I again objected that this could not be the case, since people still remember the event as it occurred in their own lives. He said that the nonphysical aspects of these people would have experienced both timelines, one after the other, and thus would remember the event. It would feel like a déjà vu rather than a clear memory.

I still did not agree with his hypothesis, but I had to admit that ETs, with complete command over travel across time, could probably do this. I argued that it was more likely that the ETs secretly retrieved the crash debris later, including the bodies of their fallen comrades. If they could secretly abduct people in the middle of the night, they certainly could retrieve a crashed saucer wherever it may be stored.

Many questions remained, but neither my monitor nor I planned to do any more work on the problem for inclusion in this book. We simply wanted to test whether or not the incident actually occurred. It did. ETs can make mistakes and crash, and the Roswell Incident was, and is, real.

My monitor also informed me that two other remote viewers accurately targeted the Roswell Incident recently, further corroborating the evidence I obtained. How it was resolved, and why there seems to be no physical evidence of the event now, are still not known to either of us. We could pursue this information, but quite frankly, the event seems small compared to other things that we are looking into. I am certain that remote-viewing historians will one day record all of the details of the event. With time and effort, remote viewers can also solve the question of whether or not the ETs manipulated time to resolve the event or something less dramatic concealed the physical evidence of the crash.

POSTSCRIPT ON ROSWELL

The front page of the national edition of the *New York Times* on the 18th of September, 1994, carried a story about the Roswell Incident. The article quotes government sources who say that they can now speak the truth about that mysterious event. They claim that the original government story about the crashed vehicle being a weather balloon was fraudulent, but the truth is that a different sort of balloon created the crash, and this balloon was used in a top secret defense project called Mogul. The story goes on to say how the scientists involved in the project did not like keeping silent about their activities during the years that the flying saucer story took hold and blossomed, but that they did not want to hide it any longer.

I do not claim that the defense project called Mogul never existed. Nor am I claiming that a government balloon did not crash. But there is a big difference between a balloon of whatever complexity and an alien spaceship complete with dead and live ET inhabitants, and it is not possible to confuse a humanoid ET with a fragment of a balloon. Government disinformation attempts are no longer useful, if they ever were. If the government cannot be truthful about the Roswell Incident, then silence would be the best course to take until such time as the real truth can be told.

CHAPTER 25

Earth's Future Environment

At times it seems that in almost every situation in which one encounters a serious discussion of extraterrestrials, the subject of human abuse of the Earth's environment arises. It was a central feature of ET-originating information in *Abduction* by John Mack (1994). This was one of the reasons I sent my mind into the near future of our planet.

A few years earlier, some remote viewers had completed a privately funded project investigating the long-term impact of the increasingly ominous ozone hole. In the course of that project, the remote viewers had witnessed a radical alteration of planetary vegetation patterns and a tremendous reduction in the human population around the globe. They also reported seeing large domes located in desertlike environments that housed and protected much of the remaining human population on the planet. The details of this project are still proprietary and have not been released to the public by those who funded the investigation. I mention only some of the general highlights to help readers understand the impetus that drove my monitor and me to look at the long-term future of the planetary environment. As with all other Type 4 data situations, I was given no advance warning that the current session was to target our planet's future ecosystem.

Date: 28 July 1994
Place: Ann Arbor, Michigan
Data: Type 4, remotely monitored
Target coordinates: 6121/6026

The preliminaries indicated a target associated with movement and artificial structures.

CB: "I am getting black and white colors. For textures, I get pavementlike. The temperatures seem cool to cold. I am tasting tar, whatever that tastes like, and I am smelling burning and hearing fire. I get initial impressions of explosive energy and rapid motion. I have to put this down as an aesthetic impression: I do not like the feeling of this place. I have to AOL the sense of death."

MONITOR: "Stage Three."

CB: "Sketching now. I have two diagonal arrows coming down hard into something that is circular or ellipse-shaped, flat. I am also getting an AOL of the concept of 'end of Earth.' "

MONITOR: "Put that as an AOL and move to Stage Four."

CB: "OK. To begin, I am probing the Stage Three arrows. In the Stage Four matrix, I am getting something shiny, fast, polished, grey, and steellike. There are beings in the structure, rooms. It seems like a container of some type. It is hitting the surface fast. Following the impact, there is burning, pain, fire, and rocks. But the beings are not dead. They are suffering, however."

MONITOR: "Move to Stage Six and construct a timeline. Locate the current time on the timeline, and then locate the event that you are seeing. Using the incrementation method, estimate the time difference."

CB: *Long pause.* "The event occurs about 290 years from now."
   My monitor has me execute a movement exercise that places me one-quarter of the way into the future between the current time and the time of the event. We label that point A on the timeline. He then gives me another movement exercise that places me at that time over the target area so that I can get some sensory impressions.
   "This place stinks. Things are different now. I really get the

strong AOL of Los Angeles. The place has lots of pollution, the taste of burning, sounds of motors, horns, city noise. This is an urban environment that seems to go on forever, spiderlike, flat, expansive. I get the sense of great wastefulness."

MONITOR: "Move further up the timeline, halfway between our session's current time and the time of the crisis. Label this point B. Dump your sensory impressions of this point in the matrix."

CB: "The textures are dirtlike, black. The temperatures are warm to hot. I taste both burning and an ammonia taste. This place is stinky like all hell. This is desolate here. I detect no life. The place is dead."

MONITOR: "Go back to point A on the timeline and cue on the five leading causes of change as one advances up the timeline from the current time to point A. Execute these maneuvers as quickly as possible. Keep up the pace."

CB: "OK, I am cuing on concept number one. I am AOLing Los Angeles, and I definitely get the sense of being on Earth. I get lots of people, activity, even commotion. This is a city, a very busy city. Moving to concept number two. There are too many people for the system to handle. There are health problems, high levels of sickness, pollution-caused diseases, and new diseases. Concept number three: food is a problem. There seems to be trouble with resources to house and feed the populace. This place has overrun its carrying capacity. Concept number four: radiation is a problem as well. The plants and animals are not growing well. There is increased desertification in many places. There is widespread destruction of trees, seaweed, plankton, fish, the general food chain, and the overall growing environment. Concept number five: energy is causing problems and becoming a catch-22 situation. The worse the other problems, the more energy is needed to solve the problems, which causes further problems. The ultimate problem is human disregard for life other than its own. The ecosystem is breaking down everywhere, and people are still trying to maintain a failing system."

MONITOR: "Try to locate any type of key sanctuary that may exist at point A on the timeline."

CB: *I receive an immediate and strong tug toward the newly refined signal.* "I am getting a beige color, a texture of cement, temperatures of cool to warm, and a whirring sound. As for dimensions, I am picking up square, angular, deep, hollow, and very big."

My monitor then moves me one thousand feet above the sanctuary. From this perspective, I draw a sketch of some type of walled structure or compound. It seems to extend deep underground and connects to numerous tunnels. The structure has a cover or top as well. I then move to an itemization stage of SRV and begin dumping detailed data.

"This structure is very complex. It is like an underground city, newly constructed. It is walled. The walls are used to protect the inhabitants from invaders. I am getting a *Mad Max* scenario as an AOL from the signal line. I am writing that down now. In general, the place is built to protect the inhabitants from the elements and from roaming gangs or tribes.

"I am also detecting a personal reaction from myself. I get the sense that I dislike the plan in some way. I will dump that as an aesthetic impression. Moving on. . . . The elite chose who could live in the structure. It seems like the Greys had some involvement in the structure plan and/or choice of inhabitants. They had gene pool considerations. Other than that, humans chose who lived in the sanctuary. The others outside were left to tough it out."

At this point, my monitor had to end the session in order to attend to some personal business. He told me that the target was "Earth ecosystem/near- to midterm." However, since I was already accurately bilocated on the target, and since there seemed to be a number of loose ends that needed to be tied up, I decided to continue the session solo. The remainder of the data for this target were collected under solo conditions.

I moved back to the compound that I found at point A on the timeline. Probing this compound, I found it to be newly constructed. The idea behind the compound's construction was long-term survival of choice humans. This is where I got the distasteful aesthetic impression of elite control. It seemed that the choice of some people as being better than others was made on grounds that may have had much to do with wealth and influence.

I returned to my initial sketch of the arrows and the ellipse that indicated some sort of crash. I needed to figure out what was going on at that point in time—the very fact that I knew my unconscious picked it told me it was important to the target cue.

Probing the ellipse shape that appeared on my Stage 3 sketch, I detected rooms, and a sense of a complex command chamber. I got the clear impression of a military operations center. Some of the inhabitants wore uniforms, there were weapons, and there were sheltered bunkers of some sort.

Probing the two diagonal arrows that pointed into the ellipse, I found a structure that was a vehicle, a spaceship, actually, on a collision course with the command chamber. There was a sense of desperation from within the vehicle.

Shifting back to the compound at the crisis point, I found that the facility was old, and that the populace was considering coming out onto the surface of the planet. The arrows into the ellipse identified an attempted landing that went awry, not an attack. There was no sense of hostility. The crash impacted on the top, or near, the compound. The inhabitants of the ship were members of the group in the compound.

The crash point on the timeline (290 years into the future from the 1994 session time) was the point of emergence of the sanctuary's habitants onto the planet's surface. I cued on the mission of the ship before the attempted landing, and I found that the occupants were making observations of the planet's biosphere status. They were in an orbital ship, measuring everything from various surface conditions, atmospheric readings, radiation levels, and water purity. Apparently, the crash did not seriously damage the compound. Moving slightly forward in time, I noticed that the compound was unchanged, and that the inhabitants were involved in their normal daily affairs.

DISCUSSION

This session was a powerful one for me. A half hour after the session, I was resting on my bed still somewhat bilocated, and I got the impression that it will be the Martians and the Greys who will teach us how to live underground. But the Martians will be the ones who actually live with us, holding our hands, until the re-emergence to the planet's surface.

The implications of this session on my view of evolving humanity is great. It seems that we really are heading into tough times. Using an incremental method in SRV, I can estimate that point A on the timeline was approximately seventy-two years into the future. This would mean that around the year 2065, the situation will have deteriorated to the point where the construction of the sanctuary structure is necessary. A string of disasters follows, one after another, into the following century. By 2150, chaos is either widespread or imminent.

I do not know what ultimately reduces the size of the human population; war, disease, and hunger are the obvious guesses. But it seems clear that the human population will not stay at its current (and near future) high levels. Three hundred years from now humans re-emerge to the surface of the planet again to begin their reconstruction of the surface biosphere. Other remote viewers have observed Greys collecting a large, and possibly complete, collection of biological samples from across the planet. I can only guess and hope that such material will be used to re-cover the globe with life.

Interestingly, I now wonder if the long period living underground will accustom humans to this type of housing environment. Robert Monroe has reported that he witnessed (during some of his out-of-body experiences) humans on Earth a thousand years in the future who do *not* live on the surface of the planet (Monroe 1994 and 1985). Rather, they live underground and regularly visit the surface, where a pristine natural environment robust with life exists. At this point, I do not know whether this will turn out to be the case. I have not yet remote-viewed this far into the future.

It is sad that humans will have to endure such hardship as a price for learning how to respect the value of life other than their own. Do not misunderstand me in this respect. I am fully aware that many humans are greatly distraught over the way that the environment is treated. Many people are actively trying to influence the course of human affairs to avoid the future that many people intuitively sense our world is approaching. But these sensitive people are too few, and their efforts are not adequate, given the scope of the global condition. Indeed, I cannot see anything that could be done to change the unfortunate characteristics of our near-term future. It seems that our race is destined to learn the hard way, doomed to destroy our world in order to understand the delicate complexity of its value to us.

It will be wonderful to see the day when the collective mentality of humans begins to look significantly away from the greed of its isolated members and struggles to find meaning in existence that connects individuals to the remainder of creation. This moment will be the genesis of a brighter future for our people. Suffering may continue for some time, but with such a change in our collective awareness, a much brighter future, filled with meaning for the existence of all life, will certainly arise. At least, this is my hope.

CHAPTER 26

How the Greys Are Organized within the Federation

The Greys have a complex organizational structure, as this chapter reveals. I am a political scientist, and it is normal for me to analyze any group of individuals in terms of how they operate collectively. Typical social scientific problems deal with beings who communicate by telephone, committee meetings, newspapers, television, and radio. Governance under such circumstances is relatively straightforward. A society forms some institution, like a parliament, that identifies itself as the government, and a social scientist who wants to study that society begins by watching that government. You gather your data through normal communication channels, such as publications and interviews with the leaders of the government, or opinion surveys and voting analyses of the populace.

But with the Greys, the situation is not so straightforward. Since I did not know exactly where to begin, I decided to add a general target to the project's long list of targets that would seek to identify basic aspects of the organization of Grey society. The specific target was "Greys/current governance-organization." I had no expectations, and I did not know how I would begin analyzing any data that I received. Fortunately, as is characteristic of remote viewing, the general cue allowed my unconscious to organize material in a way that ultimately made a great deal of sense to my conscious mind.

This session was monitored, and, as always, I was given no information regarding the target until after the session was completed. The data presented here are of sufficient complexity that readers can easily comprehend the usefulness of using a monitor for such general targets. During the session, my unconscious put me in the correct location to obtain the data, but my monitor had to do a lot of navigating once I was there in order to extract all of the bits and pieces that were required.

Date: 30 July 1994
Place: Ann Arbor, Michigan
Data: Type 4, remotely monitored
Target coordinates: 1443/7114

The preliminaries immediately indicated that the target was either in another time or dimension. There was lots of blue and white light associated with the target, together with ideas of expansive, endless, infinite.

CB: *In my Stage 3 sketch, I draw a strong light and I AOL the Federation headquarters. I am given a movement exercise that places me at the center of the target. I move into Stage 4.* "I am in a large structure, and it seems very familiar. I am having the strong AOL from the signal line that this is the Federation headquarters. I am inside the building now.

"I am now standing right in front of that Buddha-like fellow with whom I have interacted before. He definitely knows that I am here."

MONITOR: "Cue on the concept of the relationship between the Federation and the Greys."

CB: "He is telling me that I already know. I should not doubt so much what the Greys say. He seems happy, but also a bit annoyed at our slow progress." *At this point I express a bit of exasperation, thinking that I am writing and remote-viewing as fast as I competently can, given my other responsibilities. The Buddha fellow does not seem to mind this mild outburst.*

MONITOR: "Cue on the idea of line of command."

CB: "OK. Doing that now. The Greys are members of the Federation. They are very well respected members, in fact. The entire Federation is involved at the current time with the project to help the Greys. They have more than paid their dues. They deserve their chance. They have even helped humans tremendously, both now in the real things that matter—spiritual evolution—and in the past. They are a very well behaved species, and they will evolve into beautiful and even more valuable members and leaders of the Federation."

I move into Stage 6 of SRV. "That wise old Grey that I have worked with before has just joined the conversation."

I execute an advanced SRV technique in which I probe for the structure of the Grey organization with regard to the Federation. [See Figure 1.] In the sketch, I identify the command structure of three separate groups of Greys, all with relation to the Federation. The upper levels of these three separate groups of Greys are the authorities that represent their own respective populaces. In general, the Greys that interact with humans at the current time can be categorized as primitive, modern, and advanced. Each group works with the Federation independently of the other two groups.

Figure 1: Organization of the Greys

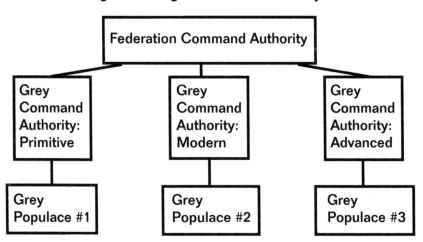

"I am getting the sense that there are also a few small renegade Grey groups that do not work directly with the Federation. But the three big groups do work with the Federation.

"I also get the sense that the three basic Grey groups are separated by evolutionary levels. That is, they are the same Greys, but they originate from different points in time. Yet all three populaces are interacting simultaneously here on Earth."

MONITOR: "Courtney, probe the Grey populace number one."

CB: "Wow! That sure is an AI! Let me put that down in the matrix. There are *huge* numbers of Greys in this category. They are relatively primitive. They have large eyes, and they feel few surface emotions. They have a hive mentality. They operate extremely large ships, and very few remain on their homeworld."

MONITOR: "Probe the idea of governance for this populace."

CB: "They actually have a hierarchy for governance. There are high- to low-ranking members."

MONITOR: "Probe the Grey populace number two."

CB: "Again, there are large numbers. These guys have some physical differences from the number one group. These Greys sometimes work with advanced humans and other beings, unlike the first group."

MONITOR: "Probe the idea of governance for this populace."

CB: "The governance is more 'full' than with the first group. There is telepathic communication on all matters. Consensus is reached. There still is no idea of individual dissent requiring a vote. Yet there still is a hierarchy, as with the first group."

MONITOR: "Probe the Grey populace number three."

CB: "Large numbers again, but not quite as many as for the first and second populaces. They are also substantially different in character from the other groups. They have gained or regained some emotional flexibility. They are not the same as humans, but they are close."

MONITOR: "Probe the idea of governance for this populace."

CB: "These folks have sufficient individuality that they have differences of opinion and uniqueness of personality. They actually vote in some telepathic way. However, the bias is still to move toward a consensus."

MONITOR: "Probe the Federation membership requirements and responsibilities associated with the first Grey populace."

CB: "They applied for membership. They were given provisional status at first. They worked for a very long time as helpful members on many projects. They applied for help with their own project. The seeds of their project began with their observation of other advanced Federation species. They saw the opportunity to gain the abilities, skills, and evolutionary potential of these other beings. But they wanted something unique. They value themselves and feel that some critical aspect, combined with some diverse other genes, will yield a biological machine that they will feel comfortable with and that will be particularly valuable to themselves and others. Federation officials agreed to their plan. The entire plan was first submitted to the Federation, and then accepted. Earth was chosen as the central gene pool bank for the project."

At this point in the session, my monitor had to leave to attend to other matters. Since I was already bilocated, I decided to continue the session solo.

I returned to face the Buddha-like Federation fellow, and I probed for information regarding the command structure of the Federation. I was informed that there is an elaborate organization. There are real physical meetings. It is not unlike the United Nations on Earth. It is not a strong central authority, but an organized collaboration of species, homeworlds, and groups. The influence of the Federation does *not* extend throughout the entire galaxy. Other galaxy organizations do exist, but there is only limited interaction with them. The situation is much like *Star Trek: The Next Generation*. The Federation is expanding, slowly. The Federation eventually wants human, and specifically Earth, participation. This is the goal. *Earth needs a world government for this to occur, however. This is the single most important criterion for membership in the Federation. The Federation will not deal with planetary factions.* It is against Federation rules to do so, except when the groups—such as the Greys—are separated by time.

Again, facing the Buddha-like fellow, I probed on the concept of guidance. I was told that things are urgent. Knowledgeable humans must teach about Federation matters. It is not long before events on Earth accelerate. The link to the Federation needs to be established in a rudimentary way now so that contact can be maintained during the years of struggle, death, and turmoil. I was specifically informed that the Federation will not solve our problems, but it can and will give guidance if asked.

Continuing with the signal, I clearly discerned that we need to apply for Federation membership soon. The application can be initiated in many ways. The world government simply needs to make the *mental* decision to want Federation membership, and the Federation will take care of the remainder. The Federation will openly act then.

I thanked the Buddha-like fellow and the Grey who seemed to assist me earlier, and then ended the session.

I called up my monitor later that day and found out that the target was "Greys/current governance-organization."

DISCUSSION

The key to understanding the Grey social structure is to accept the fact that linear time does not exist in subspace. Linear time is one of a number of structuring aspects of our existence in the physical universe. But it makes sense only within this overall physical structure. Extensive remote viewing over many years by many trained viewers has demonstrated that *linear time does not exist outside of the physical world.* All events happen simultaneously. This makes little sense to most of us on Earth, where time is sequential.

Remote viewers can pierce time as easily as space, and to remote-view events at different points in time gives the sensation of actually being at those places at those times. Thus, if a remote viewer were to observe a boat race, the event would be perceived in its own real time. Indeed, the remote viewing would become part of the event, and another remote viewer watching the same event could perceive the subspace aspect of the first viewer as he or she also watched the event.

We now know that the Greys have long had a technological ability to move both their ships and bodies through time. Although I do not understand the principles involved, it is possible that they do so by temporarily shifting their presence from physical space to subspace, making the movement in both time and space, and finally re-emerging at a new time and location in the physical universe. Remote viewers have no data to suggest that the ETs have the ability to defeat the speed-of-light limitation in the physical universe without first leaving the physical universe. Thus, from a strictly physical perspective, they encounter the same relativistic laws that we do.

Since Grey society, linearly understood, has had the technological ability to move in time for a long while, it makes sense that we could be visited by groups of Greys who are from different time periods. However, what is perhaps most interesting is that evolution or time separations seem to act on physical beings similarly to geographical separations. The relatively primitive Greys do not seem to enjoy working with the modern Greys, and neither of the two earlier Grey groups work intimately with the advanced group. In a very real sense, the three levels of Greys are much like separate nationalities among humans. We often find that our common national experiences are not easily understood by humans from other nations. There are language and custom differences, and we sometimes simply view the world differently than our fellows across the border. Similarly, Greys from one time period do not fully understand Greys from another time period, and when they interact in the same time period, as they currently do near Earth, they keep their distance from one another. This is true despite the fact that the earlier Greys are trying to evolve to be like the later Greys.

I can perhaps place this into a broader perspective that would add clarity to the discussion. Remote viewers now know that Earth is also being visited by what appears to be future humans who also have the technological ability to move through time. Thus, it is entirely possible that present-day humans could witness the flight of a ship that is being piloted by future versions of themselves. If such a ship were to land in my back yard, and my future self were to emerge from the ship's hatch, I am not certain that I would know what to do. Depending on how far into the future my future self came from, I am not certain I would feel comfortable interacting with him (if he was a him).

Without question, more research needs to be done with regard to all of these matters. But we now have an initial understanding of the complexity involved with our interactions with both the Greys and the Federation, and we now must expand our understanding not only of the interaction of cultures separated by physical distance, but by time. Comprehending the time-simultaneity of galactic interactions will be perhaps our greatest intellectual challenge.

CHAPTER 27

Buddha

In the early period of these investigations into the ET enigma, at the same time that I suggested to my monitor that I would like to use SRV to interview Jesus, we decided to include the personalities of Guru Dev and Buddha on the list as well. I have presented the results of all my contacts with these beings in previous chapters, except for that with Buddha. Buddha himself was an unknown to me. Quite honestly, I wanted to include him so that I could obtain discussions with a cross-section of great spiritual leaders in my analyses. But I did not know much about Buddha. Although I have read some books that discuss Buddha and Buddhism, such sources never left me with much of a solid idea of who he is or what he actually did on Earth. In short, he was an unknown to me, and I really wanted him on my list of targets only because of his reputation as a great spiritual leader. (I am a bit embarrassed to admit the shallowness of the original idea.)

Nonetheless, Buddha was not insulted by my lack of knowledge of him. Moreover, from what I do know of Buddhism, many Buddhists may find my interaction with him to be far from unexpected. He has always been somewhat of a mystery personality in my understanding. Indeed, in the session that I present in this chapter, he remains so, but at the same time, he teaches more about the meaning of life than I could ever imagine learning without assistance in one lifetime. Simply, the SRV session upon which this chapter is based was the most beautiful experience I have ever had while remote viewing.

This session was conducted blind. As readers will see, my monitor was caught a bit off guard by this session while it was unfolding. Buddha is truly a master teacher. There was no way that we could possibly have been prepared for what happened at midnight one summer in Michigan.

Date: 30 July 1994
Place: Ann Arbor, Michigan
Data: Type 4, remotely monitored
Target coordinates: 1842/3355

The preliminaries indicated that the target was in a hot place.

CB: "I am getting black, white, and beige colors. Whatever this is, it seems flat, wide, circular, and very expansive." *I draw a simple sketch in my Stage 3 that consists of a large oval.* "Moving to Stage Four. I am getting blacks and whites, hots and colds, lots of contrasts in light and temperature. Again, this is circular, fast, energetic, and now I can perceive that this thing is swirling. I get the sense of burning, like smoke. But the smoke is in a vortex. It may be something other than smoke, but it is in a vortex. I at least get that much."

MONITOR: "Cue on the ideas of vortex and inside."

CB: "OK. Doing that now. WOW! What an aesthetic impression! I get the idea of something that I heard of long ago in which Buddha expanded his consciousness to be really large, like a mile in diameter, and then he shrunk his consciousness to be really small, like on the molecular level. I am feeling like my consciousness is currently *really* expanded, in the stretched sense, like Buddha's was. I also get really strong energetics, very powerful.

"I get the sense that I am in a very large area, almost like a galaxy, and that small parts are spinning around. I am feeling stretched thin, but not uncomfortably.

"I sense that I am supposed to get something out of the big picture here. I am also getting something like an allegorical overlay. I have the initial image of the Buddha-like fellow in the Federation headquarters. The current situation reminds me of when I entered his mind and he showed me a picture of the galaxy.

"But this current scene is not the galaxy. This is very different in a weird sort of way. The idea here is of creation. Life is not yet formed here. The essence of life is present, but it is not manifested yet. It will come, however.

"What should I do now?"

MONITOR: "I am not sure. Do you have any ideas?"

CB: "Let me ask my unconscious. I will cue on the idea of suggestion in the matrix."

MONITOR: "Follow your intuitions. We have never done this before."

CB: "I am getting the idea of cuing on God. I will do that now. Hold on. . . . I am now getting the idea of moving to a personality that I know, Jesus. I will cue on Jesus. OK. I have Jesus now. I am cuing on the question of where I am right now.

"The signal is clear. I am being told that I am where life began at its initial point. I am going to cue on why I am here.

"I am being informed that I need to know the reason for life. This is a very powerful signal. I am now cuing on the question of what I should do now.

"Jesus tells me that I should release. I should go into the vortex. It has been good knowing you, buddy. I am going inside.

"Oh my gosh. What an AI! I am inside now. God spent an eternity as a point source. His evolution reached a point of change in which he could not bear the loneliness of isolation. His only recourse was to re-create himself throughout an infinitude, thereby initiating a sequence of creating new gods, new hims, beings to care for him, and for him to care for. He loves us because we ended a loneliness that was beyond any ability to describe in words.

"He will never allow the demise of his creations, since it would send him back to the point-source past, and that was his prison. The future is to forever expand.

"Oh, wow. This is just overpowering! I just had an experience of an explosive shift. All of time is created. There are galaxies, infinite diversity, all expanding. There is great joy, great joy in God's new existence. Great Joy!

"I am now asking if there is more that I need to know. I am a bit shaken. I am not so sure I can go on much longer.

"Jesus is telling me that this is the end. I can rest now. He says good-bye. Where did you send me that time?"

MONITOR: "Courtney, it was not me who sent you. The target was Buddha."

CB: "Buddha?!"

Immediately after this monitored session ended, I continued with a brief solo session.

I cued on the Federation Buddha-like fellow. I felt a warm sense from him.

While remaining within the protocols of SRV, I asked him if he was Buddha. He told me to leave it as a question. He said there was no need to give a specific answer. He also suggested that others should find him in their own way. But my conscious mind was clear on the matter. He was one and the same. He was Buddha, and I had been interacting with him all along and did not know it.

I asked him if I should put the information about him in my book. He told me to decide for myself. I would include it, though I could never do justice with my words to the bliss of his being.

DISCUSSION

Buddha did not want to tell me directly who he was; he wanted me to *experience* the answer to my question. Nonetheless, as incredible as it may seem, Buddha sits on the Federation Council that helps monitor the affairs of humans on Earth. To this day, he watches over us.

Buddha wanted me to release myself so that I would expand and embrace, sense, experience God's creative essence, and indeed, the moment of creation. It was this intimate experiential knowledge that led me to the certainty that Buddha and my Federation teacher/friend were the same. From his guidance came a general principle: *Revelation (i.e., being told something) is the infantile route to knowledge. Experience is the mature route.*

Apparently, Buddha felt that I needed to know the reason for life itself. This chapter's results are the counterpart to the results presented earlier in the chapter on God. Based on my experiences drawn from this session, life exists because God wanted to create life. God's motivation was to end his own loneliness. The sense of loneliness that I experienced in this session was the deepest, most penetrating pure sense of the concept that I could ever imagine experiencing. Moreover, the joy that God experienced when he created time, physical matter, and us, was similarly the purest, most wonderful joy that could ever be.

I now understand what is meant when it is said that we are made in God's image. It does not mean that God has hands and feet. It means that he feels as we feel, or perhaps more appropriately, we feel as he feels. Emotion, the rich blood of experience, is Godly. Now I know why the Greys want so desperately to evolve with physical bodies and brains that can experience emotion, especially love. They want to move toward their ultimate evolutionary destiny, union with God.

Based on all of my experience remote viewing, I am certain that the Greys with whom humans often interact know much about the actual scientific structure of God. But these same Greys know about God more as data. They want and need to experience God, and that they have not yet done.

How differently I now look at the Grey's evolutionary fight to advance in their lives. They have technology that is the wonder of the galaxy, yet they are not satisfied with this accomplishment. They have the ability to move transparently through time and space, yet their travels do not satisfy them. They are completely telepathic, yet they seek more of their minds. They have everything that causes us to wonder in awe, yet they would move heaven and Earth for a genetic piece of us. They want to *feel* again. And, indeed, God wants to feel *through* them again.

It is clearer now than it ever was for me. We, and all of life, are made from the substance of God. Since we are parts of God, he experiences life through us, and we experience life because it is God's nature to do so. Moreover, I used to fear that God may collapse again back into his original point source one day, and then my true subspace existence would end forever. My fear was based on the idea that infinity is a long time for virtually anything to happen. I now no longer fear that. God would not collapse himself back to the point source, because that would bring him back to the state of loneliness through which he lived an eternity. Since I would not want to lose my existence, neither would God. God would not voluntarily return to his own hell.

Rather, the future is for God to expand forever, for him to increase his manifest complexity throughout infinity. God will expand as we expand, and that perhaps answers the question as to why we humans feel driven to grow rather than to stagnate. It is God's nature to grow and to expand in his own evolution. Since we are made of his substance, it is our nature as well. Deeply embedded in our consciousness is the same God-dread of isolation and loneliness that creation, growth, and expansion destroyed. We are, quite literally, chips off the old block. We are in God's image. We do as he does. We seek to understand the as yet unknown possibilities of existence. Indeed, our own struggle to survive is God's struggle to Be.

CHAPTER 28

Martian Culture on Earth

As a social scientist, one of my natural interests is culture. One of the first things that I mentioned to my monitor when I met him for the first time was that I thought remote viewing could be used by social scientists to study other cultures. At the time, it seemed to me that his military orientation focused his attention too narrowly. Military remote viewers seemed more concerned about the logistics of the ET operations—who was flying what, where, and how—than the societies from which these operations originated.

Following my normal interests as a social scientist, one of the first targets that I included on the list of targets for this book project was the current Martian culture. I wanted to find out what Martian society is like now. We knew Martians flew ships, and therefore they had achieved a high level of technological sophistication. But human society in the United States has high-tech types as well, and knowing this would tell a visitor very little about the lives of young people in today's inner cities in America. Thus, I needed more than information about pilots and geneticists. I needed to know about the common folk—who are they, where are they, and what are they doing?

Late in July 1994, my monitor assigned me the Martian culture target. The session was blind, of course, and in my case, there was an especially important reason for this. It was important that I enter the session unguided and unaware that the target had anything to do with Martian culture.

Interestingly, when the session was over, my monitor told me that many of the other remote viewers who work with him had independently corroborated the data that I collected. I have since seen the written data from these independently collected probes. He then told me many details regarding the logistics of how the Martians I observed were being supported, and we ended the session both commenting about how nicely the interests of a logistics expert and a social scientist fit together.

Date: 31 July 1994
Place: Ann Arbor, Michigan
Data: Type 4, remotely monitored
Target coordinates: 4731/8279

The preliminaries indicated that my initial approach to the target involved a moving, hard, artificial structure.

CB: "The colors are grey and steellike. The textures are shiny, polished, even luminous and glossy. The temperatures are roasting. I am getting something moving very fast, together with the idea of linearity—perhaps of movement—and high energetics."

In my Stage 3 sketch, I draw what appears to be a fast-moving ET ship. I move into Stage 4, where I collect a small but sufficient amount of data to confirm that the structure is an ET ship. I execute a movement exercise that places me fifty feet above the ship's destination.

"I have a complex structure below me. The colors are browns, tans, and reds, mostly. The textures are rough as well as smooth. The temperature is warm. I smell some burning, smoke. There is quite a bit of noise at this location as well. The general dimensions of the place are wide and flat. There are lots of curvy lines below me, and quite a few angles appear as well."

I make a new Stage 3 sketch, which appears to be of a group of one-story dwelling units that are located adjacent to an open area.

"Moving on to Stage Four again. The geometrical pattern below is quite intricate. I get the sense that this location is somewhat crude, even primitive. There are lots of beings here, and I am getting the AOL from the signal line that this is like an old village from long ago, but I am not saying that this is in the past. It is just like that type of thing. The structures are dwelling units, and the textures of the structures are woodlike. Moving in closer now. Yes, these are relatively primitive dwelling units.

"The beings themselves are OK, in the sense that they are calm. Life seems to be slow and peaceful here, again in the sense that there is no immediate disaster happening.

"Looking over the place more generally, this is a moderately sized town with humanoid beings. Getting a bit closer to the beings, they do not seem to be ordinary humans, however. Something is a bit weird—hold on. . . . They are shorter, but there seems to be more to it."

I execute another movement exercise, which places me two hundred feet above what turns out to be a town. I again return to Stage 4, where I go inside one of the structures. I let my unconscious determine which of the structures is important to enter.

"I am in a large room now. This may be a large one-room structure. At the very least, this one room dominates the structure. This is a warehouse of some kind. There are things stored here—boxes. Wooden boxes. The boxes have staples in them. I am examining one right now. These are enclosed shipping crates. Do you want me to go inside the box?"

MONITOR: "Go inside."

CB: "There are medicine, drugs, medical items, things like that. These things were brought here in some kind of relief effort, sort of like what the U.S. was trying to do for Somalia." *My monitor has me return to an earlier Stage 3 sketch and probe what is inside some of the other living units.*

"These are simple people that live in these structures. They live simple lives. They have families, children. There is cooking going on. The medical stuff in the first structure seems more advanced than these people's culture."

I move into Stage 6, where I draw a schematic representation of the flight of the original ET ship from its starting point to its destination point. Then I follow the ship backward to its starting point.

"This is a very modern facility. There are beings here, and they are wearing uniforms. The quality of the facility is similar to that used by the Martian survivors in the caverns in New Mexico."

I return to the ship's destination point, placing myself one thousand feet above the terrain.

"There is lots of vegetation here, almost a jungle. There are mountains nearby, perhaps an old volcano. There are also lots of clearings. It is almost like an area near a rain forest, but not a pristine one by any means.

"The dwelling units are short and squat, simple. They are located in a forested environment, isolated from humans, it seems. I am getting the strong AOL from the signal line that this is a refugee camp of Martians. This seems like a southern location, perhaps Latin America."

MONITOR: "Just put that all in the proper place in the matrix and keep on going. Remember you are in Stage Four. Let your unconscious solve this problem."

CB: "I am probing the idea of a camp/village. Yep. These are refugees. They are a bit concerned, but they know that things are working as best as possible, at least for now. They know who they are, but maybe not everything. There may be something missing from their memory, done purposefully in order to help them assimilate with their surroundings.

"The facial features seem South American Indian. The dwelling units have been chosen in order to enhance a disguise. At some point in time, something will trigger in their minds, and their memories will come back in full. This is actually fascinating. They have their own culture hidden in their minds, and they do not know it."

At this point, my monitor was nearly ready to end the session. He had me move to a Stage 6 worksheet, where I sketched the flag of the country within which the village was located. Then he ended the session and told me that the target was "Martians/current culture."

DISCUSSION

My monitor told me that other remote-viewing sessions done by others, both monitored and solo, corroborated what I had just observed. My session, in fact, confirmed what he already knew: the general location of the village.

Readers should be clear that I am not saying that an entire race of people in Latin America are Martians. Rather, there is a small group of Martians (perhaps hundreds—I do not know the exact number) who are cleverly concealed and integrated into the larger, human population.

Following this session, my first response was that we should go to our travel agents and purchase tickets to visit this location right away. But I was reminded that this place was currently suffering from political unrest. Organized and well-armed drug runners could also pose a threat to Americans traveling through their territory looking for Martians. I remembered I had a family, and we decided to wait for an opportunity to visit the Martian community safely. One thing was certain. The Martians picked a perfect location. They were safe from visitors, even those who knew who they were and where they were. Though they lived on the surface of our planet, we could not reach them.

These Martians that I observed in this village are not the only Martians that exist. It seems as if these Martians are being supported by other Martians who are more technologically oriented. My own analysis suggests that the village Martians are volunteers in a project designed to save as much as possible of their original culture. The origins of this culture probably date back to the destruction of Mars's ecosystem and their rescue by the Greys. Their full cultural heritage is being stored in their minds while they are waiting for some triggering signal that will release their memories.

The plan is amazingly well structured. The Martians need to survive. But what exactly needs survival? I see no reason why the subspace aspects of the Martians could not be reborn as humans. But then they would be humans, and all that was originally Martian would be lost. What needs to survive are the memories of Martianhood. The original culture needs to be preserved. Additionally, some aspects of their original genetics may also be preserved in their current bodies. But their bodies on Earth are already far from the original Martian variety by necessity. The greater gravity of Earth would necessitate this. The culture could survive more or less intact if it was placed in hidden memories within a populace that was not given access to a robust human environment.

Placing these Martians in the middle of, say, New York City would probably cause them to be psychologically swamped with so much new human stimuli that any plan to resurrect their Martian identities at a future time might be seriously compromised. But a rural village where contact with normal humans was more limited would be the perfect location for this group. In my own mind, I now view this village as a cultural bank.

Thus, it seems that the plan of the Martians to migrate to Earth is quite complex. On one hand, there is a cultural repository, a bank of history and identity that will serve to preserve the knowledge of who the Martians are and from where they came. On the other hand, we now know that there are other Martians who are involved in other aspects of the migration, and these other Martians do everything from flying advanced spaceships for the support of the Earth-bound Martians to working on genetics problems.

Data from another SRV session (not presented here) indicate that there will one day be a large flotilla of Martian ships carrying refugees from Mars to Earth. Many of these Martians are now located in underground shelters on Mars and are eagerly awaiting the signal to embark on their journey. Also, many of these Martians are not highly technological. Their preferred lifestyles (including their desired climate) seem to resemble closely those of the village Martians that I observed on Earth in this session.

Thus, it would make sense that the time of cultural awakening for the village Martians who currently live on Earth would be just prior to the arrival of their compatriots from Mars. *Earth-based village Martians would be perfect tutors for the other newly arriving Martians.* They would be able to tell them much about how to survive on this planet. They would be a cultural oasis on a foreign world.

I have decided not to publish the location of this Martian village. The situation in this respect is much different from that of the Martian base in New Mexico that seems to be located underneath the mountain Santa Fe Baldy. In the case of the base, I am hoping that the publication of its location will encourage the Martians to seek an open dialogue with human governmental leaders.

With the case of the Martian village, however, I do not want to compromise the activities of the Martians. Publication of the location of the village could seriously jeopardize the safety of the villagers. Moreover, the primary idea behind the construction of the village is to preserve Martian culture. The last thing they need is a bunch of humans invading their delicately constructed haven, or making it impossible for the technologically advanced Martians to continue their relief and supply operations to the village.

My hope is that I will one day be able to persuade a senior political figure from the United States or elsewhere that a discreet visit to the Martian village would be helpful in enhancing diplomatic relations between humans and Martians. It should be possible to obtain an agreement for the visit from the high-tech Martians on Mars or at Santa Fe.

CHAPTER 29

Reality Check #2

This chapter describes the second of two SRV sessions that used a calibration target. The first such session had the Oval Office in the White House as its target. As I described in that chapter, calibration targets are used to check the accuracy of the SRV protocols. Such targets can be easily verified, and they turn an SRV session into a sort of remote-viewing tune-up.

This was one of the last targets that I remote-viewed while staying in Ann Arbor, Michigan, during the summer of 1994. I had done a great deal of remote viewing during the two weeks prior to this session, experiencing everything from the joy of the Martians' final departure as their flotilla left for Earth, to the intense love of God as he created the universe. My nervous system had absorbed an immense amount of data from my unconscious, and it was getting to be time for a rest.

My monitor said he had a simple target. The only information he gave was that it was in the past. He did not tell me whether it was a place, an event, a person, or anything else. I was in for a surprise, in a session that would tell us much about the interaction between the waking state of consciousness and the unconscious mind.

Date: 31 July 1994
Place: Ann Arbor, Michigan
Data: Type 4, remotely monitored
Target coordinates: 3102/2137

The preliminaries indicated that the target was on dry land. There were artificial things on the land, but they did not seem like buildings.

CB: "I am getting colors like beige, brown, tan. The textures are dirt, dry, but some wetness. The temperatures are warm."

I draw a Stage 3 sketch of a large open area with movement all throughout it. I am far above the target. I execute a movement exercise to place me in the center of the target site.

"I am getting the same colors and textures as before, but now I am hearing voices."

At one thousand feet above the target, I make a similar Stage 3 sketch as before, and I move into Stage 4.

"I am perceiving humans here. I am getting some kind of AOL of the signal line, but it does not want to come through. I am making an overlay of a soccer game, but I know this is not correct, so I will put this down as simply an AOL. Something is weird here.

"I am going down to the surface of the site now. There are humans here. They are wearing costumes of some sort. There are lots of colors around, primary colors. There is also a great deal of activity here, fast activity. But again, something is really weird. I can't perceive any of the emotions of the people. It is a total blank. It is as if there is absolutely no intensity to their emotionality. But they are moving around as if there is a great deal of concern here.

"I am smelling something burning now. I taste sweat, and hear yelling. This is a very confused mess down here. There is no one coherent signal. There are many people doing confused and different things. But nothing constructive is going on. I am now getting the AOL of the signal line that this is a battle of some kind."

At this point, my monitor suggests I begin probing around for anything else.

"Well, I still see the humans in their costumes. All of the colors are here. There is still that confusion and intensity everywhere. The trouble is that the people do not know what they are doing here. There is some disagreement and fighting, and things are all messed up. Actually, these people are experiencing everything in terms of confused emotions.

"I am now going to try to probe their purpose for this activity. I am getting the sense of an explosion. These people are trying to accomplish something against all odds. I get the sense that many of these people cannot or will not succeed. But they do not see that. It is like some type of campaign where there are many deluded individuals who think something glorious will happen, but it will not.

"My mind is really resisting these data. Whatever this is, my mind does not want to see it. I am shifting a bit forward in time so that I can get out of the confusion. Hold on. . . . This was a battle. Look at all of these dead people. They are everywhere. They are all dead. Bodies on a field, weapons, uniforms.

"I am getting the strong AOL of the signal line that this involves the Confederacy. This was a major battle in a war. I know this scene. This is the largest battle of the Civil War, where more people were killed than in any other battle. This place is Gettysburg."

MONITOR: "OK, Courtney. Stop the session. The target is the 'Battle of Gettysburg.' "

DISCUSSION

Besides successfully identifying a calibration target, this session caused my unconscious mind to cooperate with my conscious mind in order to protect me from the direct experience of the battlefield. Interestingly, I could not clearly identify the scene until after the battle was over and the emotions had subsided. It was only then that my unconscious and conscious minds opened their dialogue with a sufficiently wide aperture that I could see and feel everything at the site. In short, it was more than my mind was ready for in the beginning of this session, and I had to wait until my mind found a location in time and space that conveyed the needed information without overloading my nervous system. It must be emphasized that all of this occurred automatically, without any conscious decisions on my part.

One last point for the historians. The actual Battle of Gettysburg more than lives up to its reputation. It was truly a terrible thing, and words can hardly do it justice. Honestly, one just has to be there to understand it, and it would be well worth the effort for historians to revisit the battle using SRV.

CHAPTER 30

Santa Fe Baldy

Underneath Santa Fe Baldy, a mountain in New Mexico not far from Santa Fe, is a Martian base that serves as a center for their planetary operations. Much of the data supporting this claim have been presented in previous chapters. However, early in my research, I decided to do something special to back it up. Thus, I decided to add a target to the list that specifically identified the mountain itself rather than only the Martians. This would eliminate the possibility that I and others have been looking at a mountain that looks very much like—but is not—Santa Fe Baldy.

As with all of my monitored sessions conducted for this book, this one was done blind under Type 4 conditions. However, when reviewing these data, one must remember that the unconscious is highly intelligent. It does not merely obey simple requests for data. It knows intimately what the remote viewer needs to know for the project at hand. It will always direct the remote viewer to the correct answer, even if that answer is not the one that the conscious mind of the viewer thinks is right. In my case, I wanted to know if Santa Fe Baldy is the actual mountain under which there is a base. But my unconscious led me to this information by way of a fascinating data stream that placed the fact of geographical precision within the broader context of human and Martian activity associated with this particular base, both currently and in the near future.

Date: 2 August 1994
Place: Ann Arbor, Michigan
Data: Type 4, remotely monitored
Target coordinates: 4471/3621

The preliminaries indicated that the target was associated with dry land and artificial structures.

CB: "I am perceiving red, brown, and beige colors. The textures are rough, and some smooth. The temperature is somewhat cool, and I do not smell anything right now."

My Stage 3 sketch shows an area that seems to have a number of structures on it. I then execute a movement exercise that places me one hundred feet above the target.

"I am now over a building with a circular shape."

In another Stage 3 sketch, I draw the outline of a building that many remote viewers have identified in relation to the Martians. [Other sessions, not reported here, have indicated this building will be the location of high-level discussions among humans about Martians.]

My monitor then tells me to move directly to Stage 6, where I draw a small representation of the building in the center of a piece of paper. I then probe around the building to get an idea of the surrounding environment. There are other smaller buildings near the circular structure. I find a forest and a mountain a number of miles to the east of the cluster of buildings. A population center lies to the west. The mountain seems closely related to the circular structure.

"I am now moving into the large circular building. Inside, I find a spacious room set up like an office. The walls of the office are curving, following the curvature of the building. There is a door in the room that leads to a hallway where there are many small offices. Returning to the large room, I notice that there is an area for presentations, sort of like the front of an auditorium. There are inclined chairs arranged in rows.

"I am changing my focus to the people in the room. I notice that they are wearing ordinary business suits. The building itself seems to be associated with some sort of manufacturing interest.

The premises appear to be research-oriented in connection with some product of manufacture, since I detect that lots of activities occur within the structure. But I do get the sense that at least some of the activity is related to computer software."

I locate on a timeline the present day and four important points in the future. Probing the separate time points, I find at the second point (approximately two years from the present) a large rectangular modern structure under construction near the original circular building. Some of the smaller buildings have been demolished to make room for this larger building. At all of the later time points on the timeline, the large rectangular structure dominates the scene.

Moving back to the present, I find that there is a plan for the construction of the larger building.

MONITOR: "Courtney, return to the large structure again in the future and find out what the people in the structure are doing.

CB: "OK. Hold on. . . . I am now in the large rectangular structure. Inside the building, there are computer terminals, laboratory benches, extensive wiring, and laboratory equipment. This is a scientific research lab connected to private commercial interests, and I sense that the subject under investigation involves genetics and biotechnology."

My monitor has me follow one of the male workers as he leaves the structure at the end of a workday. He has to pass through the front gates of a large compound, where a guard stands. I follow the worker as he drives to the larger population center to the west and home. The environment along the route gives me the strong AOL of the signal line as being near Santa Fe.

MONITOR: "Return again to the large structure and find out what is on each of the major floors."

CB: "My unconscious is pulling me to the basement."

MONITOR: "Follow it."

CB: "Oh my. There are some dead human bodies wearing minimal amounts of multicolored clothes here. This is not a good place. I am continuing my probing. The deaths of these people are to remain a secret. They died in the line of duty. Others needed to warehouse the bodies to get them out of the way quickly. These others will eventually dispose of these bodies. Safety seems to be an issue here."

MONITOR: "Probe what you mean by 'line of duty.' "

CB: "These were scientists in the facility. They died as a result of some of their work."

MONITOR: "What were they doing?"

CB: "They were engaged in dangerous scientific experiments. They knew the risks, but did the things anyway. They were not supervised from any outside superagency. The activity was in-house and private, even secret."

MONITOR: "What was the activity?"

CB: "It involved radiation and genetic mutations of organisms. These folks were killed by products or by-products of their own labs. It seems as if they lacked proper safety supervision.

"My mind is now pulling me to the east, to that mountain."

MONITOR: "Follow it."

CB: "I am in the mountain now. There are caverns. And there are beings in the caverns. This is the Martian base, but there are changes. There are wheeled vehicles in the caverns now. The place is modern, but not too modern. There are no ET ships right now. There is a tunnel. It goes west and it connects with the surface, and it is camouflaged on the outside. The tunnel is being used as an air vent for now, but it is very big. Vehicles could go through it.

"There are many workers in here. These folks look a lot like humans. In fact, they *are* humans! They are wearing single-piece white uniforms.

"Wow, this is interesting. I will put that down as an AI. It seems that some of the dead folks in the basement of the large structure outside seem to be connected to these mountain chambers.

"This place is more active than I have ever seen it. There is lots of construction. It is like the workers are expanding the facility.

"I am moving forward in time now. In the near future, the facility is ready but empty, except for supervisor types. A little further on, the place is absolutely filled with Martians, refugees. Some of them are dirty. There are a lot of voices in here now, a real babble with lots of commotion and goings-on. The Martians are packing these chambers with children and adults who are filled with strong emotions, hope, fear, excitement.

"The Martians want to get out to the surface. They are really happy and excited!"

MONITOR: "Courtney, go back to the current time and see if the people in the circular structure know anything of what is going to happen. Then move into the future to probe when there is an awareness of change."

CB: "At the current time, the folks in the circular structure know nothing. By the second time point on my timeline, the government has made a decision to locate the housing site near Santa Fe. Money pours in at that time. The idea is to use an expanded Martian underground base as the processing center for the Martians as they arrive.

MONITOR: "OK, Courtney. We have enough. Let's end the session. The target was Hugh Hefner's last bachelor party at the—"

CB: "Come on. . . ."

MONITOR: "The target was 'Santa Fe Baldy (Martians/present survivors/New Mexico caverns).' "

DISCUSSION

This was my first look at where the Martian refugees will arrive and how the Martian base below Santa Fe Baldy will be transformed into an immigration-processing center. Many of these Martians seemed to be quite ordinary. Generally, they were not high-level tech types. They were mothers, children, average Martian adults, and so forth. Indeed, I got the impression that their immigration to the United States, among other places, would be similar in some socializing respects to that of many other new ethnic groups. I am certain humans will have to get used to this idea at first. But after the newness wears off, these Martians have the potential to be received as friendly neighbors.

CHAPTER 31

Official Contact with the Martians

One of the most important questions on nearly every reader's mind by now is probably how communication with the Martians will be initiated. As I have indicated in a previous chapter, humans will have official contact with the Martians before we will with the Greys, although how much later we will deal with the Greys, I do not know. An open request to meet with the Greys broadcast by the United Nations would speed things up. But contact with Martians will come first in any event, and this will do much to shift human awareness outward to the stars.

The session upon which this chapter is based was originally intended to find out how well the Martians will have succeeded in integrating themselves with human culture. This session provides an answer, and as with so many other remote-viewing experiences, it reveals a wealth of other important information. This other information hints at how the official human-Martian interaction should (or perhaps will) proceed in its first steps.

This session was conducted blind under Type 4 conditions, and it occurred after an extended break in monitored viewing, which I requested after the previous session in order to rest. Now returning to monitored remote viewing, I was recharged and eager to see what information my unconscious would bring to light.

Date: 26 September 1994
Place: Atlanta, Georgia
Data: Type 4, remotely monitored
Target coordinates: 6068/0004

The preliminaries indicated that the initial approach to the target involved a man-made structure on dry land.

CB: "I am getting browns and tans. The textures are wood and cement. The temperature is warm, indeed, very warm. I am tasting sweat, and I hear human voices. My perception is that there is something circular and flat at the target location."

I move to Stage 3 to make a quick sketch of what appears to be a circular structure with a flat roof.

MONITOR: "Move to Stage Four."

CB: "I am in the matrix now. I am now perceiving a building clearly. I am hearing voices in the building, so I am going in. There is a conversation going on here. There are humans talking. The building is circular, and I have the sense that I may have remote-viewed this structure before. Popping outside for a moment, there are trees in the area around the building. Now back inside.

"Wow! What an AI! I just focused on the people talking in the building, and they are a very high-powered bunch. This is a top-level meeting. Let me go in their minds. Hold on. . . . They are talking about the ETs.

"These folks have on military uniforms. They are generals, admirals, the very top military brass. There is also a civilian here. It looks like the president of the United States. Let me put another 'WOW!' in the AI column."

MONITOR: "Right. Dump it and move on. Keep a fast pace. You are doing fine." *My monitor has me move to Stage 6 immediately. He has me execute an SRV technique that isolates primary subject components of an observed conversation.*

CB: "The discussion is taking place on a very practical level. The principal focus is exactly how to communicate with the ETs. They are aware that consciousness can do this, but they want something more physical. Contact through consciousness started the ball rolling, but they now need something else. One of their suggestions is to use radio. They are trying to figure out how to do this.

"The Greys are not a subject of their conversation. These folks are talking about the Martians. This is a real interplanetary communications problems. They are now focusing on the radio."

I construct a Stage 6 timeline with which I can probe specific points in the future.

"OK, I have found the point at which humans have begun to talk to the Martians. I will call this the communication point. It seems that they are using radio telescopes—multiple telescopes, not just one. The scopes are all over the world.

"The humans begin by focusing the telescopes on Mars and listening. Seems like they don't pick much up. Then the humans change tactics and begin transmitting. There are multiple questions that needed to be resolved. One big question is the matter of which language to use. Then there are communication protocols to develop.

"The Greys are watching but not actively participating in all of this. They seem interested, but in a passive sense.

"The humans are also attempting to transmit to ET bases on the Moon. They are making more attempts toward Mars, however. The ETs on the Moon are being silent.

"Initially, the Martians on Mars are silent. They feel that they have been discovered and are wondering what to do and what the human reaction will be. They always knew that this day would eventually come. They feel a bit vulnerable.

"Proceeding forward in time, the Martians decide to engage in the dialogue. They definitely send a signal back, loud and clear. It seems that they use the same radio protocols that the humans initiated.

"I am a bit struck by how these Martians look. I have followed the radio signal to Mars and am there now. The Martians are humanoid, and very human-looking now. They even have hair. These particular Martians are predominantly male. They are wearing uniforms of some type, but this is no military fighting group. The Martians don't do that. Their entire defense seems to be built around secrecy, not combat. The skin of the Martians seems to be light.

"These Martians seem to be the same [in the subspace sense] as the original Martians. But they have bodies that are comparable with those of humans on Earth."

MONITOR: "Move forward in time. Where are the Martians?"

CB: "Hold on. . . . They are on Earth. They are working with their indigenous groups, such as those that migrated here earlier. They are also getting support from human governments to continue their work. Their work is now out in the open. Gosh, these Martians really look human."

MONITOR: "Where are the Greys?"

CB: "The Greys are doing their own thing. Their genetics project at this future point in time is either done or almost done. Finishing touches are all that remain. They are not yet talking directly to humans."

MONITOR: "OK, Courtney. We have what we need. The target was 'Martians/future culture.' "

DISCUSSION

The Martians with whom we will be interacting in the future will look a lot like us, perhaps indistinguishably so. Their real differences when compared with Earth-based humans will be those of culture and technology. They will also have needs that we will have to understand if we are to interact successfully with them. But they will not come to us as "little green men." Our first open interaction with an extraterrestrial culture will be comforting in at least this one superficial physical sense. These ETs will look like us.

CHAPTER 32

Lower Earth Life

The following session was given to me without any prior knowledge of the content of the target cue. The cue was "The Elementals," and the target was not on our list of targets for this book. This was one of many targets that my monitor quite regularly gave me that was off the list and totally unknown to me. He never even discussed the general subject with me. The reason for giving me such targets from time to time is to add an extra layer of protection to the data by inhibiting me from trying to guess characteristics of the target, which could lead to erroneous AOLs. In practice, however, my own level of mental discipline while remote viewing is now quite high, and my monitor and former teacher rarely worries about the quality of my data. Nonetheless, he enjoys being extra safe, and his practice of throwing unexpected targets at me does indeed keep me on my toes during a session.

The motivation behind this particular cue is that my monitor has long wondered what happened to the remainder of subspace life when humans destroy so much physical life. All of our remote-viewing efforts have targeted humanoid beings. But my monitor was concerned about the many subspace creatures that were not humanoid that he has personally observed during his own remote-viewing sessions. To him, they seemed like "Elementals." It is a term that he uses to refer to most nonhumanoid creatures. Most such creatures are typically smaller than humans, and they often have behavioral characteristics that are quite unpredictable. He did not know if these creatures had physical counterparts, or whether they ever were physical themselves (which may be the same thing). He just knew that they existed around us, and he wondered if human activity was hurting them in any way. In short, was human destructiveness to the physical environment negatively influencing the larger community of life in subspace? If this turned out to be true, then my monitor was concerned that we may be seeing only the tip of the iceberg with regard to the true magnitude of the current and continuing environmental damage to this planet.

Of course, this session was conducted under Type 4 conditions. Immediately after the preliminaries, I realized that this was a completely unexpected target. I did not know that it was not a target on the list. I just perceived that I had no prior expectations at all with regard to where my unconscious was directing me. Whatever this was, I sensed at once that I was to perceive something that I had never perceived before.

Date: 28 September 1994
Place: Atlanta, Georgia
Data: Type 4, remotely monitored
Target coordinates: 3660/1161

The preliminaries indicated that the target included an interface between dry land and a liquid.

CB: "I am getting colors like blue and white. The textures are splashy and wet. The temperature is quite cold. I am tasting something fishy, and I am smelling seawater. Wherever I am is wide, expansive, flat, and very deep." *I then draw a Stage 3 sketch that seems to include a dry piece of land next to a large body of liquid. I also begin to perceive and draw what appears to be some kind of artificial structure under the liquid.*

"Moving into Stage Four, I am still perceiving a lot of liquid. This is like a big ocean. Well, I can at least tell that I am on some planet. The planet seems familiar, but it is not a time that I have visited before. The liquid seems like water, and it is very deep.

"There is a dry piece of land nearby. I am going there now. I am probing around here. . . . It is barren. Nothing is growing on it. It is natural, but it has the sense of being heavily influenced by some civilization. The place is barren like Mars, but this is not Mars. The colors are different, and there is an atmosphere.

"I am going back to the water now. In the water there seems to be a structure of some type. It really is huge. I am going into the water now to investigate. Hold on. . . .

"The structure is quite modern, but not by ET standards. It is primarily metal, and it seems like stainless steel. There is advanced technology, more advanced than current time on Earth, but not much more advanced.

"There are beings in the structure. I am going in now to take a better look. Hmmm. These beings have normal human clothes on. Focusing on the beings themselves. These folks, both male and female, have human faces. Their eyes are small, like humans'. These are humans, it seems, but I cannot understand the setting. I do not know anyplace like this where there are humans."

"There are multiple floors in this structure. There are channels, like elevators, for vertical movement in the structure. There are also large vents that lead to the outside into the water."

MONITOR: "We stay away from analysis. Just put all the data into the matrix and move on."

CB: "The people are feeling worry, fear. This structure is like a huge, funny-shaped submarine. I am cuing on the activities of the people now. It seems like they are not happy doing what they are doing. It is drudge work. There is activity related to getting and processing food for survival. But they also have maintenance activities with regard to the structure. They work because they have little or no choice. Their work is definitely survival-related.

"There seems to be a higher purpose with regard to their work. But the trouble is that these folks may never live long enough to see it. Their situation is rather hopeless. They are working for a *next* generation, their children and their children's children."

MONITOR: "Go back to the land. What do you see?"

CB: "OK. The land cannot support life. The place is the result of a holocaust. It is barren, just dirt and rocks. The surface is natural but with artificially induced characteristics, like sterilization, everywhere. My unconscious is tugging me to go back to the structure in the water."

MONITOR: "Do it."

CB: "There is no life outside of the structure that I can perceive. The water is as barren as the land. There is nothing, nothing at all near here. I am feeling a tug to go forward in time, so I am doing that now.

"Hmmm. This place is teeming with life now, at least in the water. Life is basically all over. This is interesting. There is both subspace and physical life here. Checking again back in the original time. There was neither subspace nor physical life. This is odd. Where there is no physical life, there is no subspace life. It is like the two exist in parallel, in cooperation. One does not exist without the other, or at least the subspace life does not hang around where there is no physical life."

MONITOR: "What do you mean by subspace life? What kind of life?"

CB: "There are these small subspace animals everywhere. They seem like the ghosts of fish. They are animal ghosts or subspace aspects. There is a strong relationship between the physical animals and their subspace aspects. It is like they are all in the same herd or flock or school, depending on which animal you are talking about. When the physical environment was destroyed, both physical and subspace life suffered. When the recovery came, they both prospered again. Also, it seems that the activities of the humans in the structure had something to do with the recovery. They did not do everything, but they worked for this end and it happened."

MONITOR: "OK. Let's end. Courtney, this target was 'The Elementals.' "

CB: "What? That's not on the list! What are the elementals? You know I do not have much time for these types of curiosities anymore. The book is due at the publisher's office soon, and . . . what are the elementals?"

MONITOR: "Courtney, let me explain. It is very important that we understand the elementals. I have been seeing them for years. They are involved in all of this, and we cannot ignore them just because they are not humanoid." *My monitor goes on to explain what the elementals are, and why he is concerned about them. It takes me the remainder of the day, but I eventually realize that his concern is justified, and that I must mention these life-forms in the book. I have also obtained remote-viewing data that suggests that many of the ETs may be as concerned about these other life-forms as they are about us.*

DISCUSSION

It has been repeatedly reported in the UFO abduction literature that humans are told by the Greys that they (the Greys) cannot stand by while humans destroy both physical life and its associated "other-dimensional" life that exists nearby. Quite honestly, before now, I have never understood what was meant by this. It seems that humans are generally unaware of this broad spectrum of lower-level nonphysical life, and perhaps totally unaware of the delicate relationship that exists between the physical and nonphysical aspects of this life.

In part, human ignorance is due to the problem humans have in perceiving subspace (or nonphysical) life at all. In many circles, it is still quite controversial as to whether or not humans really have souls. Most people, when asked, will say that souls exist, but scientists have rarely tried to measure one. With this high level of confusion among our scientific intelligentsia regarding our own human subspace aspects, it is no wonder that we have not even begun to ask whether other forms of life have subspace counterparts.

Humans generally do not see themselves as caretakers of life on this planet, but rather as owners of a garden that is theirs to use. With such a view of other life, it is no wonder that the question of a subspace existence for all life, human and nonhuman, is rarely ever even raised. But I now know that the question is important to ask, and I have an intimate understanding of its answer.

Subspace life, all of it, relies on physical life. I do not know all aspects of this reliance, but I know that it exists. When we damage our physical environment, when we destroy species, or make existence difficult or painful for nonhuman life, we inhibit the ability of subspace life to prosper and evolve. Physical and subspace life live in parallel; one does not progress without the other. If we humans are to evolve into true galactic citizens, it may be that we will have to change our views of other life, all life. This may be even more difficult for us to accept than the fact that extraterrestrials exist.

CHAPTER 33

The Event That Destroyed Mars

If Mars once had a robust ecosystem, what destroyed it? Remote-viewing data from the period prior to its destruction does not indicate that the Martians themselves had the technology to destroy their own environment, let alone the planet's atmosphere. Based on data presented in previous chapters, we now know that the Greys destroyed their own homeworld through their environmentally reckless actions, and that humans appear to be following a similar route. But Mars is a different thing altogether. The collapse of that planet's environment had, from the beginning of these investigations, seemed connected to some natural disaster. A number of different viewers concluded that the disaster followed a celestial event, perhaps associated with a comet or asteroid.

Thus, my monitor and I constructed a target cue that was designed to identify the cause of the collapse of the Martian civilization. As it turned out, this remote-viewing session would be one of two final sessions we had scheduled for this book. You will recall that after compiling the initial list of targets with my monitor many months earlier, I had carefully avoided looking at the list of targets to prevent activating my conscious mind into forming opinions prior to the actual sessions. The list was long—that and my monitor's penchant for adding targets to the list on an ad hoc basis ensured that I was never tempted into thinking about the targets that were not yet assigned. But as the second-to-last target rolled around, I did consciously remember that the target identifying the collapse of the early Martian civilization had not yet been assigned (a situation which, incidentally, never occurred with any of the other targets). When the session began, the initial signal from the target indicated that this session was indeed the Martian target. Thus, this session that began under Type 4 conditions actually acquired a mix of Type 4 and Type 6 conditions.

Date: 29 September 1994
Place: Atlanta, Georgia
Data: Type 4, remotely monitored
Target coordinates: 5966/2695

The preliminaries indicated that the target involved motion and hard, natural land.

CB: "I am picking up colors of brown and beige. The textures are rocky, and the temperature is very cold. I hear tremendous wind noises, like a hurricane. I am also perceiving something circular and round, and I am AOLing the Mars disaster."

MONITOR: "Just stay in the structure and move on to Stage Three."
My Stage 3 sketch is a simple representation of two circular objects.

CB: "Moving to Stage Four, it seems like at least one of the circular objects is a planet. I am still detecting brown colors, rocky textures, and something is cold. I am also detecting strong atmospheric disturbances, particularly a swirling motion. There are beings involved with this planet, and currently they are in a state of terror. There is tremendous commotion down there. I am AOLing two things, both are of the signal line. One is something like a moon or asteroid, while the second is Mars."

MONITOR: "Courtney, go directly to Stage Six. You have the right target. It is the Mars disaster. Just stay in structure and continue."
My monitor has me sketch the planet and the smaller object. I locate Earth relative to these two objects and execute an SRV technique that allows me to

*identify the direction of the motion of the smaller object relative to Mars. I
also create a timeline of the event.*

CB: "The smaller object is irregularly shaped, lopsided. It has an extremely thin atmosphere, measurable only on the molecular level. This object passed through the edge of the larger planet's atmosphere. The planet's atmosphere was relatively thick, and the object pushed through the stratosphere and went on. I will call this area where the object passed through the atmosphere the intersection area. The object did not crash on the surface of the planet.

"I am probing the planet now. Initially, there was not much damage to the atmosphere. High turbulence existed in the area near the intersection. Elsewhere, little happened, just a trembling through the entire planet's atmosphere. There are higher levels of turbulence as one gets closer to the intersection area.

"Following the initial turbulence, the object caused a circular ripple to form in the atmosphere, much like the way a stone dropping into a pool of water forms an expanding circular wave outward. This ripple grew into an atmospheric tidal wave.

"There was no initial impact on the surface environment resulting from the atmospheric disturbances. It was not like an earthquake in which everything gets destroyed right away.

"The circular ripple traveled through the atmosphere and met at the other end of the planet, bounced off or passed through itself, and then returned to the intersection area. It then met again as a circular ripple closing in at the intersection area and repeated the bounce off or pass through phenomenon, on and on, producing an oscillation, vibrating in its own way like a guitar string. It developed a resonance. The resonance became the primary driver of the atmospheric conditions, swamping all other sources of influence, such as heat from the sun. Apparently, the gravity was not sufficiently strong to dampen the oscillations quickly. Thus, they continued for a long while.

"The beings on the planet were affected gradually. All weather patterns changed. The conditions on the planet slowly began to deteriorate. Food became a problem, since crops could not grow. Rain became a problem. There were both floods and droughts initially.

"The atmosphere was thick enough to breathe for quite some

time, but the continuous rippling gradually threw the atmosphere off into space. The gravity of the planet was not capable of counteracting the kinetic energy of the rippling."

MONITOR: "OK, Courtney. We have enough. The target was 'Martians/civilization collapse (event).' This is fascinating. We never could have guessed at this process. This will open up an entire new area of research among scientists who study turbulence and fluid dynamics in planetary atmospheres. I can imagine a mathematical model emerging from this."

DISCUSSION

The process I witnessed was absolutely fascinating. It is something that simply could not have happened on Earth, because of its greater gravity, which would have quickly dampened the atmospheric ripples created by the passing asteroid or comet. But on Mars, large ripples formed over a long time. There was time for the Martians to realize something was seriously wrong with their planet and for the Federation to dispatch a rescue team composed of Greys.

CHAPTER 34

The Future Culture of Earth

This chapter presents the data from the final monitored remote-viewing session used in these investigations. While the target was on the original list of targets that my monitor and I generated for the book, the current session was different from the last regarding Mars in that I did not remember that this target was on the list. The results of this session took me by surprise, and, as it turned out, pleasantly as well. In fact, until I finished this session, I was beginning to have a fairly pessimistic outlook for the future of humans on this planet. I even wondered why the ETs were putting so much effort into helping us if we were headed for species suicide anyway. Happily, I now know there is a reason for their efforts.

Date: 30 September 1994
Place: Atlanta, Georgia
Data: Type 4, remotely monitored
Target coordinates: 4395/0241

The preliminaries through Stage 2 indicated that the target involved dry land, liquid, and artificial structures. Also, there was the immediate impression of movement across time.

CB: *My Stage 3 sketch resembles a rotating globe. I perform an SRV movement exercise that directly places me at the target site. I find myself in a dense and complicated environment.* "Wherever I am is very complex. I get the sense of a complex ecosystem. There are beings here, humanoids of some type. The place is like a jungle with lots of vegetation. Everything is connected to or dependent on everything else, like a circuit board, or a jungle habitat. Indeed, this feels like a well-balanced jungle. I am AOLing the Garden of Eden, but it is not the Garden of Eden. It just feels like that. This place is well taken care of."

I move to Stage 6, where I construct and investigate a timeline. I locate the target time and three other important points between that target time and the day of the session.

"There seem to be about three hundred years between today and the target time. At target time, I can perceive that the humanoids are definitely humans. They are wearing normal clothes, and they seem to have jobs related to the environment.

"The first intermediate point in time marks serious large-scale environmental degeneration. The third is the point of genesis for regeneration of the environment. Hold on as I check the other points. . . .

"At target time, there is the beginning of a robust ecosystem, in the sense that it is gaining a self-sustaining foothold. The initial sense of complexity that I got in this session was related to the literal complexity of the vegetation in the ecosystem.

"Still at target time, the humans do not seem to have aboveground shelters. They are walking about without gas-powered vehicles. They are observing, not foraging for food. Their mentality is one of preservation, not exploitation. In the minds of these folks is the sense that they have gotten through the worst part of something, and now they are confident that they can rebuild their planet. Before this, it seemed that there were always doubts."

MONITOR: "Cue on the concept of biodiversity."

CB: "There is not nearly as much as today, but more than, say, a hundred years before target time. The key emphasis of their efforts at target time is to develop complexity in an interactive planetary ecosystem."

MONITOR: "Cue on the concept of Federation/interaction."

CB: "The interaction between these humans and the Federation is throughout the entire period from the current session time to the target time. The Federation seems to watch and offer guidance, but it's not here to get the humans out of trouble. Humans have to do this for themselves."

MONITOR: "Cue on the idea of representation."

CB: "At target time as well as earlier, there is a subspace human representation. These representatives no longer seem to be physical. But soon after the current session time, there is a dialogue between physical humans and the Federation. As time progresses, physical humans see themselves as operatives, representatives, perhaps cooperatives, who are closely associated with the Federation. I get the idea that humans evolve into types that are like the original Adam and Eve managerial species in the early Earth genetic uplift project. But this time, the uplift managers evolved from this planet rather than having originated from elsewhere.

"By target time, humans have evolved from territorial owners to caretakers, and even the idea of gardeners makes sense, but on a planetary level. There is still not a complete regeneration of the planet. But there are gardens, or patches of robust life, and the patches will spread out later.

"The ecosystems are open generally. But there are greenhouses for the not-yet-introduced species."

MONITOR: "Cue on human habitats."

CB: "Hold on. . . . At the current session time, there are no special habitats. At the first intermediate point on the timeline, there is trouble, but people are only beginning to think about the idea of creating special sanctuaries. At the next point, a *Mad Max* scenario begins, and humans are beginning to scramble. Desertification becomes extreme. There is still life, but it is mostly a desertlike, or at best a savannalike, environment. At the third point, the special sanctuary-type human habitats are fully operational."

MONITOR: "Cue on the human protocols for Federation dialogue."

CB: "OK. Doing that now. . . . There is nothing special. The Federation knows English as well as the other human languages. They will not expect anything unusual. The Federation will make all necessary communication links. The wait is for humans to signal their readiness."

MONITOR: "Directly probe the idea of Federation assistance."

CB: "There will be informal assistance only. Humans must get themselves out of this situation. The Federation input is passive in the watching sense, but active in the sense of giving humans a goal for which to work. They will not bail us out. The need is not for humans to be a dependent species, but a mature and helpful species. This maturity only comes through experience."

MONITOR: "OK. I guess we will have to wait to find out what that destiny is until the next project. Let's end the session. Courtney, now we have hope again. The target is 'Earth/future culture.' "

CB: "It is? I totally forgot about that target; it really has been long since I worked on the list. 'Earth/future culture' . . . looks like my unconscious knew what I needed to know with regard to this book. We *are* going to have a second chance!"

DISCUSSION

We humans *will* change. We will witness the destruction of our own planetary home. There will be a long period of great hardship for all of humanity. Yet our loss will not leave us empty. Collectively, we will learn from our mistakes, and our species will rise again, the next time not as terminators of life, but as caretakers and preservers of our world. But before discussing the rebirth of humanity, let me focus a moment on some of the logic behind our coming decline.

Some readers may challenge the notion that the Earth must become a planet of dust before it is again reborn with wiser human caretakers. While the remote-viewing data clearly indicate this as one phase on the future trajectory of our planet, logic can lead to the same conclusion given our improved knowledge of some of the driving factors of the human mind. The problem is not just that there are too many humans for the planet to sustain given its theoretically limiting carrying capacity. The planet could sustain many more humans than it does now. The trouble is that our own human nature is such that nearly all humans will continue to desire a physical lifestyle that is equivalent to at least the better-off elite of this planet. On average, we will continue to exploit our natural resources to satisfy short-term physical desires. Our weak connection between the conscious mind and its subspace aspect leaves most of us no alternative but to constantly chase happiness through our physical senses. The vast majority of humankind will continue their struggle to enhance their physical well-being without limits, until the planet's carrying capacity is so severely strained that it crashes, thereby destroying much but not all of the human population with it. It will be at this point that the so-called *Mad Max* scenario develops, and humans begin to think in terms of surviving the planetary environmental siege.

The Federation will not save us. If it did not stop the Greys from self-destruction, why should it stop humans? But now look at what good has come from the current direction of the Greys' evolution due to their own past experiences. From such hardships grows greatness, and humans have an important future destiny.

I realize that most readers will see this near-future scenario as bleak. But the truth is that our future as a species is bright. I encourage readers not to be so blinded by the seriousness of our near-term challenges such that the glory that awaits our species just beyond these challenges is not recognized. Throughout nearly all of my remote viewing in this book, I focused on the problems inherent in our current situation. These problems include our environmental troubles, the weakness in the mind-body connection for our species, and the interaction between ourselves, the Federation, the Greys, and the Martians. Indeed, everything that I have witnessed somehow relates to the general idea of many species trying to resolve their collective problems. This has been a necessary but nearsighted limitation of my study.

The current session, however, bids us to look farther into the future. Sometime around the year 2300, we humans will recover significantly from the problems we brought upon ourselves. As a group, we will be the better for it, having become more mature as a species. We will turn our attention outward to the world and universe that surround us, and we will work to help life we encounter which struggles as we once struggled. Our mental outlook will be softer. We will have learned how to love in a broadly defined sense.

While I have not sent my mind further into the future than the next three hundred years, I can only assume that such a new and wiser human species will not sit around and rot. In previous sessions, the Federation folks indicated to me that they wanted humans to join the organization as full members, and to help the Federation continue its expansion throughout the galaxy. The flavor of consciousness that I perceived from the beings that I watched in the current session was not that of warrior-type galactic explorers, but of celebrants of life. When future humans are gentle beings, but not passive, and destructiveness is no longer a trait of their psychology, our voluntary bond with the Federation will be complete.

I suspect that these future humans of the year 2300 are the prototypes of even more advanced humans that I trust will soar through our galaxy by the year 3000. I can only barely imagine the role that we will play in the galactic dramas that will unfold before us as we interact with other species, and as we help each other through the troubles that will always challenge the evolution of life.

It humbles me to realize to how few places and times I have sent my mind. I have no idea what we humans will be involved with two thousand years into the future, or even beyond that. Will we eventually be leaders in the Federation? Will we eventually help the Federation succeed in extending its influence throughout the remainder of the Milky Way? Will we even eventually extend our hands to realms of our universe that exist in other galaxies?

All of my remote-viewing efforts have demonstrated to me beyond any shadow of a doubt that we indeed are *more* than our physical bodies. Our combined physical and subspace personalities need never end their evolutionary march forward, and thus I feel pure joy as I consider the unknown possibilities of existence. Just beyond our challenges and hardships, we face an unbounded future of excitement and wonder that extends quite literally throughout time with no end. God was not cheap in his gift of life.

In the next few years, we humans will learn how to interact with the Martians. After that, we will begin interacting openly with other species, including the Greys. Eventually, we will venture off this planet using our own ships, after we have brought its life force back to a state of health.[7] Beyond that, I do not know what we will do. But I will be there, as will everyone else, and collectively, one day, we will all discover what comes next. For now, it is best to face our life experiences in manageably small bites. Now is the time for us to purge ourselves of fears and reluctances and look toward Mars. This is the next step in our species' evolution toward greater galactic involvement. We must begin now to talk to the Martians.

---

7. It is possible that we could use remote-viewing data to determine the nature of future events and then change our behavior to create a new timeline where those events do not occur. In other words, it is possible for the present and the future to interact. But given the current genetic dysfunctionality of humans—especially the weak mind-body connection—I doubt that we have the ability to avert the grim ecological disaster that is soon to be upon us.

PART III

A HUMAN APPROACH TO GALACTIC POLITICS

CHAPTER 35

Training the Diplomats

If humans are to enter the realm of galactic diplomacy, it is absolutely essential for us to recognize that there are (at least) two forms of life—physical and subspace. *We are composite beings.* Our physical forms are inhabited by subspace life-forms. Physical life-forms are temporary creatures, in the sense that they eventually die. The subspace aspects of these life-forms persist, apparently forever. Our human "souls" are simply our subspace selves that existed before we became physical, and will continue to live after our physical bodies decay and slough off.

Advanced sentient races understand all of this, and they actively communicate across the physical-subspace divide, often using technology to bridge the gap between the physical and subspace realms. The advanced races that I have observed can perceive both realms simultaneously using their own physical and subspace nervous systems. Humans, perhaps due to our own unique genetic structure, do not normally have this ability as a natural characteristic, though we can be trained to develop it. Competent and professional training is not inexpensive, and I advise only those with clear data gathering and communication needs to obtain all of the training that is available today. Human galactic diplomats would be ideal candidates for such training.

Scientists and historians are other groups that would benefit from this training.

In this chapter, I outline a complete course in physical-subspace interactive communication. The course consists of three distinct parts. The first part of the course involves the learning of a specific meditation technique, and I recommend it as a minimal entry level of training for anyone who wants to proceed further. I also recommend it for individuals who want to expand their own personal growth in the direction of advanced ETs and future humans. I know of no precautions that need to be taken with regard to this initial level of training. Indeed, I suspect it would be a welcome trend if children sought out this level of training generally, finding it to be more "cool" than dueling with laser blasters.

The second level of training is specifically designed to introduce students to altered states of awareness. The U.S. military considered it a prerequisite for training in remote viewing. This second level of training involves working with the Hemisync technology developed by the Monroe Institute.

The third level of training is formal instruction in scientific remote viewing.

It is very important for all readers to understand that none of the companies or groups that I identify here either formally or informally endorse the complete training program outlined below. Nor do they advertise their products as useful to investigating UFOs and extraterrestrials. This training program is a product of my own research. None of these companies or groups is in the business of training people to be galactic diplomats. But my experience indicates that what they do teach can be spliced together in an effective manner to achieve this end. But fundamentally, these are *my* thoughts on the matter, not those of the groups that do the actual training.

The parts of the course of study that I outline below tremendously complement each other. But the parts must be practiced independently of one another, without any blending. Each part accomplishes something different, and the totality of what is accomplished is what is important. In my view, it is equally important for all readers to recognize that the complete course of study is not risk-free. It is entirely possible for *some* individuals to be adversely influenced by obtaining the training that I outline below, and I do not know how common or rare this could be. Thus, this total course is designed to be followed by individuals who are supported by institutions that offer proper psychological oversight in all matters related to each student's development. In the absence of such oversight, psychological problems could result. This training opens one up to all sorts of activities and areas of awareness that are not typical of the standard human set of experiences in a lifetime. For example, some students simply may become seriously disturbed when they realize what a telepathically received thought feels like. Without proper guidance, the student may develop a paranoia, perhaps wondering if all of his or her thoughts are actually telepathic manipulations from invisible beings. People have to be ready for this, and after they receive the training, they should be observed closely to allow properly trained supervisors to offer counseling at particularly sensitive moments.

In a very real sense, consider this description of a training program like a page in an encyclopedia that describes the formula for gunpowder. Anyone can obtain this formula, and in the absence of supervision, there will always be some people who will try to mix gunpowder in their basement and inadvertently blow their heads off. The encyclopedia is not responsible for these misadventures, since the knowledge itself cannot be, and should not be, removed from the public domain if we are to continue to live in a free society. Similarly, anyone pursuing the course of study that I outline here should have responsible supervision. Any group, company, or institution encouraging or requiring this course as a function of employment simply must add the cost of supervision to the cost of the overall program. This supervision is generally not supplied by the groups that actually do the training. Trainees and their employers must supply this.

Any individuals following this course of study do so at their own risk. I am not a trained psychiatrist, nor do I have any formal training in matters that would allow me to discern characteristics in an individual's psychology that would make such a course of study dangerous to his or her mental well-being. Nonetheless, I have structured this course with the idea of making it as safe as possible to the best of my knowledge, and I know of no one on this planet at the current time who has explored these matters more systematically than I.

When I developed the course, I did so at my own risk. I had no psychiatrist to observe my progress. But on the other hand, I was very careful to grow incrementally with regard to my consciousness. I first learned Transcendental Meditation, which gave me the base of a mind familiar with its own subspace complexity, and I followed this up with completing my training in the TM-Sidhi Program before experiencing any of the products available from the Monroe Institute, and before learning SRV. Yet even with this gradual growth, I found that my experiences shook me to the core of my being. In only two years, all of the belief façade that structured my view of the world collapsed. I learned that we were not alone in the universe, and that nonphysical beings shared this dimensional reality with me. I learned that ET civilizations rose and fell in my own planetary neighborhood, and that some traveled through time with the ease that I walk across the street. I had to reformulate my understanding of God and all religion. It is impossible to relate to you how much I had to adapt and grow in order to confront the realities that were opening to me.

Thus, my approach to growth in consciousness is conservative, and this program is structured to allow a gradual experience in personal development. This safe route is also the most effective route. The course begins with safely and directly experiencing your own subspace existence and continues through an experiential awareness of concepts such as telepathy and out-of-body states, and finishes with a professional program in SRV. As a final note, while I have structured this course sequentially, with each part following the next, I do not know if this is absolutely necessary. What I suspect is really important is that students complete all three parts within a reasonable period of time, even if, say, the training in SRV occurs first.

The Course for Galactic Diplomats

Part I

Students are encouraged to learn Transcendental Meditation (TM) and the TM-Sidhi Program. There are many forms of meditation, but few of them actually have the meditators directly experience their own nonphysical selves. Some activities that are described as meditation are simply mental imaginations, or more ominously, stressful practices that can lead to a state of unhappiness. TM, and its advanced version, the TM-Sidhi Program, are precisely designed to place individuals into a state of direct awareness of their total selves. The practice is truly stress-free, and it has a variety of scientifically well-documented positive side-effects, such as improved mental balance, greater satisfaction with life, improved physiology, and improved intuition.

TM teachers are highly trained, and the course is identical regardless of who the teacher is or where it is taught. The standardization of instruction and course availability are essential ingredients to any larger program in which consistency is a primary concern.

The TM organization may not endorse my suggestion of using TM to help individuals evolve into galactic citizens. Rather, *my personal observations* indicate to me that people who practice TM have a flavor to their consciousness that is much like that of advanced ETs and future humans. In public lectures, TM teachers traditionally focus on the physiological benefits of Transcendental Meditation. This is probably a direct result of their experiences with human resistance to more direct guidance regarding nonphysical life. It is easier to tell people in our society that their blood pressure will improve with the practice of TM than it is to tell them that they will soon become aware of their own souls.

Nonetheless, the writings of Maharishi Mahesh Yogi are very clear on these matters. According to Maharishi, all beings, whether physical or nonphysical, inhabit a realm of existence called the *relative*. There are a variety of levels to this relative realm. Extrapolating from Maharishi's perspective, both the physical and subspace levels of existence reside within the relative realm. Maharishi also identifies a field that he calls the absolute, from which all things in the relative levels of existence originate. Moreover, he consistently refers to the need to unify the two separate aspects (both relative and absolute) of each person's existence. If TM teachers do not emphasize this during instruction, it is not a fault of theirs, nor are they hiding anything. It is necessary for everyone to learn about one's total self by experiencing this totality directly. TM is an experiential practice. Since most people's ordinary perceptions are limited to their physical senses, talking about one's subspace and absolute aspects makes no sense to most people independent of the meditative experience.

With TM, one "bookends" the day by meditating morning and evening for twenty minutes. While it is easy to meditate, the practice is delicate, and it is absolutely essential that one learn how to meditate from a competent and certified teacher. Moreover, the TM course includes follow-up sessions that ensure that the practice continues to be conducted properly.

I once had an experience in which someone asked me why she should pay for instruction in meditation when she could do it by herself for free at home. At first I did not understand her. I thought she was asking me why it is necessary to pay for this instruction at all. I told her that TM teachers do this for a living, and they must support themselves like everyone else. But later I realized that this person's question was really aimed at something quite different. She assumed that meditation is something that someone can figure out by one's self, or get out of a book. Thus, I answered her incorrectly.

Readers of this book must understand that the proper practice of meditation is not something that one can figure out independently, nor is it something that one can get out of a book. The TM course has evolved out of centuries of trial and error. TM is simple, but it is also a highly refined and delicate practice. Those attempting to creatively design their own procedure are trying to reinvent the wheel when sports cars already exist everywhere.

Readers should understand that the practice of TM and the TM-Sidhis Program is quite different from remote viewing. With remote viewing, one obtains direct knowledge from another location and/or time by exploiting a subspace mental connection with that location. This can be accomplished only because we are also subspace beings. However, with TM and the TM-Sidhis Program, there is no attempt to exploit this connection. While meditating, one experientially learns simply to be aware of one's *total* self, both subspace and physical. Thus, regular meditation can lead to the development of a personality that is complete in the sense that one is not ignorant of one's total self.

This experiential knowledge can lead to a great deal of personal satisfaction in life, because one no longer needs to chase physical pleasures through the sensations of touch, sight, smell, hearing, and taste in order to experience an inner sense of happiness, completeness, and satisfaction. The subspace realm cannot be directly accessed through these senses, so this physical chase ultimately fails. Gaining direct experience of one's other half twice a day is enormously relaxing, and it can make subsequent physical experiences more satisfying, because these experiences need no longer be laced with a hidden yearning for subspace contact.

For the purposes of this course in galactic diplomacy, students should take both TM and the TM-Sidhis Program. The courses are taught at TM centers (recently expanded and called Maharishi Vedic Universities) in most major cities and many rural areas throughout the world. They are taught conveniently in the evenings or on weekends. TM takes approximately one week for the entire course, whereas the TM-Sidhis Program takes somewhat longer.

After learning TM (no need to wait for the Sidhis to be completed), students should begin reading two books by Maharishi. The first book is *The Science of Being and the Art of Living*, which is available at all TM centers and in most bookstores. I advise reading this book slowly, perhaps consuming approximately twenty or fewer pages a day. The important points in this book can look like unimportant subtleties at first glance, and a quick reading can lead to frustrating results. The second book is Maharishi's translation and commentary of the first six chapters of the *Bhagavad-Gita*. Both of these writings contain a wealth of information on the composite nature of human existence.

Part II

This part has two sections. The first is to listen to a set of thirty-six tapes at home that are offered by the Monroe Institute, located in Faber, Virginia. The tapes are collectively called "The Gateway Experience," and my experience has been that they gradually lead the student through introductory experiences of telepathy and techniques of subspace communication and energy manipulation. I suggest listening to each tape twice, once in the morning and once in the evening. The tapes should *not* be listened to before one's morning meditation. Meditation should be the first thing in one's day. Listening to the tapes during any convenient time after meditation is fine, however.

Following this rule of listening to one tape (twice) each day, and giving allowance for rests and unavoidable circumstances, the home course should be completed in about two to three months. After that, I suggest listening to two particular tapes, each tape once every day, for another three to four weeks. There should be at least a few hours between listening to each of the two tapes to let the mind clear from one experience before entering another. Thus, a morning and afternoon (or evening) schedule for both tapes would be ideal. These two tapes are labeled "Mission 12" and "Far Reaches," and both are found among the collection of thirty-six tapes in the Gateway Experience home study course. Based on my own personal experience, repeated exposure to the Mission 12 tape helps develop one's ability to engage in telepathic communication, and repeated exposure to the Far Reaches tape helps one become intuitively familiar with the polarity shift that is so common to altered states of awareness.

Following this month of enhanced exposure to the two tapes, my recommendation is to stop using all of the home study tapes *permanently* in order to avoid the development of a tape dependency. Also, if psychological stress develops during the repeated use of these tapes, the tapes should be discontinued immediately and indefinitely, and you should take time out—with counseling—before deciding whether to proceed with the remainder of this course.

The second section of training involves actually visiting the Monroe Institute and taking the course "The Gateway Voyage." The higher levels of consciousness that are taught in this course are not available in the home study tapes. Experiencing these higher states requires the supervision of the trained staff at the institute, and I support the restriction of the sale of these advanced tapes to the general public. For example, there are enormous numbers of subspace beings who live in these higher levels of consciousness, and it is important to be exposed to these beings carefully. It is not that the beings are dangerous, but that the experience of meeting them can be rattling for some people. In general, these advanced levels of consciousness include those that one normally experiences soon after physical death. The in-residence course at the institute safely teaches one how to experience these levels well before this ultimate journey.

In terms of background reading, I suggest that students read three books by Robert Monroe. They are *Journeys Out of Body*, *Far Journeys*, and *Ultimate Journey*. All three books can be obtained in local bookstores, or they can be ordered from the Monroe Institute.

Part III

Part III of this course is composed of two sections. The first section consists of a one-week intensive course in scientific remote viewing that is taught by a qualified teacher. There are a number of former military remote viewers who are now individually teaching the Army's version of remote viewing. Also, the Farsight Institute, a school teaching scientific remote viewing in Atlanta, Georgia, can efficiently instruct large numbers of students. The Farsight Institute also offers monitoring and teacher-training programs.

Next, a qualified monitor should work with each student on a project of the student's design. The monitor will help the student work through a minimum of ten to fifteen monitored sessions. It does not matter if the sessions are remotely or locally monitored. This experience of working with a monitor for an extended number of SRV sessions produces the level of professionalism needed for future scientific and diplomatic work.

One Final Reason for Using the Entire Syllabus

When targeting clearly identifiable physical targets (such as the Oval Office in the White House), remote viewing yields a data stream of easily understandable reality. It is generally a simple matter to gather information about such targets. The unconscious mind locates itself at the target site, and one sees what one sees.

But ET targets are not always so easy. SRV data are always factual, but it often happens that a target represents a deeper idea about which one requires information, and the unconscious understands this informational need. In such situations, the unconscious often supplies data that answer the hidden agenda that exists behind the target cue. In such a situation, the remote viewer has to be able to place the data into a larger perspective. Since the unconscious has access to all information in both the physical and subspace universes, the data supplied by the unconscious could seem mysterious unless the viewer also has an expanded understanding of consciousness. This expanded understanding of consciousness is a framework within which to place all information obtained by remote viewing. To a naive viewer, such data could seem symbolic or allegorical, when in fact the unconscious is being quite literal. Thus, to understand the data, one has to have an evolved consciousness, and there is really no escape from this requirement.

CHAPTER 36

The Involvement of Human Government

Why the Government Has Been Silent

One of the most common complaints among people interested in UFOs is governmental silence on the matter. The only thing that enrages these people more is what they perceive to be attempts by governments to ridicule, repress, or dismiss UFO reports. I too went through a stage in which I felt my government was not doing its elected duty by keeping information on this subject from the people who did the voting. I have changed, however, and it may be important for me to state my perspective on this matter.

I should remind readers at the outset that I am a professor of political science. One of my specialties within the discipline is public opinion and mass behavior, which directly relates to governmental concerns regarding the subject of ETs and UFOs.

The government certainly is aware of much of the ET activity on and near this planet. Some books have been published that have focused on information released from the U.S. government using the Freedom of Information Act (for example, see Good 1987). But one need not go far to confirm that the government is aware of the fact that the ETs are active on this planet.

In my own experience, I have talked to retired senior military

officers who bluntly stated that they themselves were involved in secret and high-level data-collection activities with regard to UFOs,
and that the government was doing its best to deal with events—
but with little success. Moreover, I have talked to airline flight per-
sonnel who have told me of experiences in which commercial jets
were followed by UFOs. In some cases, the pilots were "greeted" by
high-level governmental security agents upon landing. The pilots
were debriefed, and then given strict instructions not to discuss the
matter with anyone else. These particular pilots obviously did not
follow those orders to the letter, but others do.

The government knows about the ETs, but it does not tell its citizens. Why should that be? Carefully consider the situation that
you would be in if you were the president of the United States, to
use my country's government as an example. You are aware that
extraterrestrials are invading the nation's airspace at will, without
ever asking for permission. Moreover, at least some of the ETs are
doing things to many of your nation's citizens that the citizens do
not appear to enjoy, and the government—with all of its military
and security apparatus—can do nothing to stop these things from
happening. Absolutely nothing. What would you do? Would you
go on national television to announce the arrival of the ETs? What
else could you possibly say other than, "They are here, and you can
panic at your discretion"?

You could consider revealing that the visitors have arrived, and
that the government is trying to open diplomatic relations with
them. But how long would you be able to get away with this if the
ETs did not respond?

It may not have been the correct strategy, but minimally, it was a
defensible strategy to postpone the public's large-scale awareness
of the problem until the government had a chance to try to deal
more successfully with the matter. No nation's leaders wish to announce failure if there is any hope of success.

I do not know exactly how much the government has known
over the years. But I know that the nation's leaders do not have the
full breadth of information that we have been able to collect using
remote viewing. Thus, information in this volume will be helpful to
establishing a new phase in human-ET interactions. But I see no
benefit in seriously attacking current and former governmental officials with regard to past governmental policies concerning the

ET phenomenon. Mistakes may have been made, but a perfect policy was probably never a real option given the situation.

On the other hand, I strongly believe that the time has come to seriously consider a change in the previous policy of denial. Historically, humans have always been passive with regard to interactions with ETs. We have watched the ships fly by, and some of us may have been abducted. But always the ETs came to us, and we just watched it happen. Now we have the ability to move from a passive to an active stage in studying interstellar life. With this ability must come a new grasp of our need to participate responsibly within this larger society. Just as ETs have studied our society, we can begin closely examining theirs. Moreover, educating our own public about the ETs is the first step toward establishing reciprocal diplomatic relations.

The Martians

Our governmental leaders must be closely involved in the next step of active human participation in contact with the extraterrestrials. I have already stated that this next step will involve direct physical contact with the surviving Martians, not the Greys. We will one day work with the Greys directly in a fashion that will be more satisfying to us, but that day is not today. There is nothing to stop us from opening up lines of communication with the Martians now.

First, it must be understood that any contact with the Martians must be at least partially sanctioned by leaders in our *planetary* government, however feeble this government may currently be. At a minimum, leaders of the United Nations must at least be consulted with regard to these contacts. Moreover, no attempt at contacting the Martians is likely to succeed unless the nation or organization that actually makes the contact relates all communications directly to the United Nations Security Council. All member states must be informed as soon as such a meeting can be arranged.

This is an essential ingredient to success, not a product of my own moral position. The Martians want to come to Earth. They are not going to acknowledge to humans that they exist unless they have an assurance that they are working with representatives of the entire planet. Their best defense against a volatile and often violent human species has always been silence and secrecy. They will maintain that defense unless they have a reasonable chance at attaining their goal of being accepted by the majority of humanity. They will not likely prejudice their own future success at being accepted on this planet by giving the initial impression that they are siding with one of the many nationalistic factions.

With this said, however, practicality dictates that there may be only one nation on this planet that is capable of accessing sufficient political and technological resources in order to succeed in persuading the Martians to come out of hiding. In my view, that nation is the United States of America. My remote-viewing data suggest that the initial formal contact with the Martians will involve the use of radio communications. The United States already has much of the necessary equipment for such a project, and it could enlist the assistance of other nations in supplying additional communication equipment as necessary. Radio telescopes can be aimed at both Mars and the Moon in an attempt to initiate an open dialogue with the Martians. My data suggest that any ETs based on the Moon will be silent in this dialogue, but it is probably wise to include them in the transmission circuit, since the Martians on Mars will likely consult with them regarding any Earth-originating transmission.

I suggest that the president of the United States authorize (with United Nations sanction) the transmission to Mars of an invitation to begin direct talks between Earth-based humans and the chosen representatives of the Martians. The transmission should indicate that humans are warmly receptive of the idea of working with the Martians with regard to issues of mutual concern. The message might also suggest that a prompt reply by the Martians would indicate a willingness among the Martians to become cooperative neighbors with humans, and that this would be very helpful to future relations between the two planetary cultures. This may be a polite diplomatic way of twisting their arms a bit, but it may be essential right now. The Martians have developed a habit of secrecy, and they have to be told that it is worth their effort to break this habit and to work with humans directly, and soon.

The question remains as to which president may authorize this transmission. My clear sense is that the time for the transmission is now. Thus, it would be opportune if the current president of the

United States authorizes the transmission. Yet regardless of who is working in the White House at the time of the transmission, one thing is certain. The leader who initiates successful communication between Earth-based humans and Martians will make a major impact on the development of human culture that will last for thousands of years.

In general, there is probably no single event that will influence the future of human collective evolution more than contact with an extraterrestrial civilization. Whoever is brave enough to risk doing such an extraordinary thing as beaming a transmission to Mars asking for a planetary dialogue to begin will be remembered favorably, both on Earth and across the galaxy, for a very long time. This act will finally signal to the waiting galactic community that humans are now sufficiently mature to enter the community with full membership. Such membership is earned, not awarded as a gift. What it takes to earn it is the collective maturity to accept the reality of who we are: complex beings in a universe filled with life.

Following this transmission and the Martian response, it will be necessary to begin planning to receive the Martians who are not yet on this planet. My suggestion is to offer to transform the current Martian underground base in New Mexico into a processing center. I am certain that Martian medical technology is sufficient to guarantee that no new diseases will be introduced onto this planet as a result of the Martian arrival. Indeed, if this were apt to happen, it would have happened long ago, since many Martians are already here. But this does not mean that humans should accept this knowledge passively. It will be important for human doctors to become fully aware of the intricacies of Martian physiology and psychology. Thus, we will have to process Martian refugees in much the same way that we process refugees from other Earth-based cultures.

There will be a question of citizenship. First, it must be recognized that any Martian children already born on Earth should have immediate citizenship in the countries of their birth. Martians born in the caverns underneath Santa Fe Baldy in New Mexico, for example, are automatically citizens of the United States. Moreover, their immediate relatives (such as their parents) have rights to expedited permanent residency in the United States. Similarly, the

United Nations must encourage other governments to grant citizenship to Martians born in their territories, and to extend permanent residency to the relatives of these citizens. The reality is that the United Nations will have to orchestrate the arrival of many new Martians who do not have children born on Earth, and our world governments will have to work together to develop a plan to permanently locate and accept these new and hopeful travelers in search of a home.

It is not possible to emphasize how important human behavior is in this new stage of our existence on this planet. Literally the entire galaxy is watching to see how we humans behave toward our planetary neighbors in need. Though we have not always been aware of it, we have been helped by extraterrestrials during much of our evolutionary history. The real test that we face is whether or not we are sufficiently mature to be able to look beyond ourselves and to act with compassion toward others in real need. Are we capable of participating in the single most common act of intersteller cultural interaction that has occurred across the millennia, the act of helping other species with their evolutionary struggles?

My remote viewing suggests that we are now capable of this level of altruistic behavior. With all of my heart, I hope I have not erred in my perception of this single and ultimately important piece of information.

The Greys

In my opinion, the Greys are not yet ready to work *physically* with large numbers of humans on an equal level (i.e., as one human might interact with another human). Their telepathic abilities are extremely advanced, and they have a great deal of difficulty dealing with the intensity of our emotions. Moreover, when we are around Greys, our emotions tend to move rapidly toward unrestrained panic, and so it is understandable why telepathically sensitive beings might find it unsettling to work near us in an uncontrolled environment.

But this is not to say that we should not have other types of human-initiated contact with the Greys. Indeed, it is to the benefit of both ourselves and the Greys if we begin such contact as soon as possible. My recommendation is that human diplomats should begin using SRV extensively for communicative purposes with the Greys. Moreover, I suggest that humans initiate contact with both subspace and physical Greys across all of the evolutionary types that are operating near Earth today.

Extensive experience has demonstrated that Greys can tolerate subspace human contact quite well. This means that humans have an ability to work with Greys directly via this subspace connection. This would be particularly useful, since the Greys need to have the experience of working with fully aware humans so that they can be encouraged to break their past pattern of behavior of working only with humans in a clandestine fashion. Thus, the Greys need our help in dealing with them as much as we may have benefited—and continue to benefit—by many of their efforts. In case these efforts do not come immediately to mind, let me remind readers using one such example that there exist substantial remote-viewing data from a number of viewers suggesting that the Greys are heavily involved in storing vegetation and animal genetic samples from the very environment that humans are so busy destroying. We will be very grateful for their efforts along these lines in later years when we begin to rebuild this planet using these stored genetic stocks. The relationship between humans and Greys is very complex, and we need to be both patient and constant in our efforts to improve our abilities to communicate with them more openly.

Perhaps the best way to advance communication between humans and Greys is to use SRV to ask the Greys how we might be able to assist them with the genetic project that is related to their own species' evolution. In the past, conscious and willing human help with this project has been nonexistent. Greys have had to work with humans who have little or no understanding of the complexities of subspace life.

I suspect that such attempts at communication will not yield immediate responses from the Greys, in the sense that their ships are not likely to land next to the United Nations building the moment the first human diplomat asks a Grey if we may help them. But repeated efforts are likely to produce great rewards, and we must remind ourselves that the Greys have been waiting for a long time for us to mature sufficiently to be able calmly to communicate with them.

The Galactic Federation

As with the Greys, we will need to use SRV to establish permanent human representation in the Galactic Federation. There are physical means by which humans could communicate with Federation authorities, and I am certain that these physical means will be used soon. However, SRV is necessary at this time for one very important reason: the Galactic Federation is primarily a *subspace* organization.

It is virtually impossible for only physical beings to govern a galaxy. The reason is that physical beings are temporary creatures who participate in the physical world only for brief periods of time. Moreover, much of the life of physical beings is consumed by childhood and old age, and there are actually very few years of adult productivity in the life of even a long-lived human. On the other hand, the galaxy evolves in a time frame that can only be understood as spectacularly long, relative to a single human lifespan. In order to monitor and assist the evolution of a galaxy of life, the beings must have an active memory that is much greater than, say, seventy years. The dramas of even one species' evolution often consume thousands of years, and if the Federation is involved in helping this species, then beings involved in the project would need to be around a long time. Physical beings cannot do this.

Physical life is something that all of us participate in. It has at times been called a school, where subspace beings go to learn how to be better in some way. But physical life is really much more than a school for subspace creatures. It is a very real dimension of existence. The primary difference between physical and subspace existence is simply that everyone can visit the physical way of life only temporarily, quickly making their contribution before rather abruptly leaving. But it is nonetheless an authentic way of life, however temporary our participation in it may be.

To locate the Galactic Federation within the physical universe could be suicidal to both the organization as well as many species. Given the vagaries of physical social evolution, who is to predict how physical societies may change? One day they may favor giving assistance to other societies. But physical beings may quickly change their minds if, say, their economy is not performing well one year. One cannot govern a galaxy with personalities as fickle as those that are typical of physical beings. Galactic governance requires a longer perspective, and no forms of life other than those found in subspace can yield this required longevity. Thus, it is not by accident that the Galactic Federation is a subspace organization. It could never have been otherwise.

The time will come when humans have technology that spans the physical-subspace divide. But until that time comes, we need to use our own nervous system, trained to listen through our own subspace aspects, to talk to the Federation authorities. We already started our representation in the Federation as soon as we began to use our own human consciousness to communicate with Federation authorities. Both my monitor and I were early participants in this process of human representation. Now is the time for physical human authorities to authorize such representation formally. Now is the time for the representational efforts of cosmic wanderers and explorers such as myself and my remote-viewing colleagues to give way to those of trained representatives of a human planetary government, and now is the time for the United Nations to formally sanction direct human-Federation talks.

Let there be no mistake about this. Neither the Martians, the Greys, nor the Federation authorities will do anything to *force* us to communicate with them. They are waiting for us to act first. The signal to the entire galaxy that we are a sufficiently mature species to deserve a formal voice in the community of worlds is our own ability to recognize who we are and among whom we live. We are not children anymore. We are a species with a destiny. Let us begin crossing the new frontier of that destiny proudly. Let us leave our cynicism and our fear. Let us speak, finally, to those who have waited so long and patiently for us, out there.

COMMUNICATING
WITH THE AUTHOR

Readers who would like to communicate with the author can send letters to the address below. Those who would like to know more about the availability of professional training in scientific remote viewing should write directly to the Farsight Institute at the same address.

Courtney Brown
The Farsight Institute
P.O. Box 49243
Atlanta, GA 30359

GLOSSARY OF REMOTE-VIEWING VOCABULARY AND ABBREVIATIONS

aesthetic impression (AI)—an emotional response to something that is remote-viewed. SRV requires viewers to declare (and thus rid the mind of) all AIs to prevent contamination of the data with internalized emotions.

analytic overlay (AOL)—a mental conclusion reached during a remote-viewing session. It represents mental "logical" analysis, which may or may not be correct. The protocols of scientific remote viewing require that a viewer declare (and thus rid the mind of) all analytic overlays.

AOL matching—a strong AOL that indicates to the viewer that a positive identification of a mental image has been made. These are rare.

AOL of the signal line (AOL/S)—an AOL that the remote viewer senses as directly originating from the data stream during the remote-viewing session. Such AOLs usually contain particularly relevant meaning to the target interpretation. Often correct target identification occurs as an AOL/S.

bilocation—At some point during the SRV session, the viewer's attention is so strongly directed toward the target that the viewer's awareness is split between his or her physical location and the target site.

cue—One or more words used at various stages during an SRV session in order to direct the unconscious to obtain information regarding a target or a specific aspect of a target. Initially, cues are associated with the target coordinate numbers that are used in Stage 1 of the SRV protocols. Subsequent cues are entered into the SRV matrix in order to refine the data stream, and are not used until after the viewer is securely bilocated at the target site.

emotional impact (EI)—refers to emotions that are associated with a site. They can originate from beings actually present at the site. However, a site can have an emotional impact due to previous or even future events. EIs do not refer to emotions experienced by the viewer, which are AIs as defined above.

energetics—a sense while remote viewing that a great deal of energy is being expended at the target location. This energy can be of any type, such as kinetic (as with a fast-moving ET spaceship) or radiant (as from a hot energy source, like a sun).

event—a cue that is used to locate the unconscious at a target location during the time of some significant activity.

matrix—a collection of labeled columns that are written on a piece of paper while remote viewing. Data are entered into the appropriate columns during the session.

movement exercise—an SRV procedure to place the viewer at a new location relative to the target.

session time—the date and time that a remote-viewing session takes place.

signal/signal line—the data stream that is originating from the unconscious mind during the SRV session.

SRV—scientific remote viewing.

Stages 1 through 7—separate stages of the SRV protocols. The specific stages are structured as follows:

> *Stage 1:* Stages 1 and 2 are referred to as "the preliminaries" in this book and are designed to establish initial contact with the target. The data obtained in Stage 1 are crude; for example, whether there is a man-made structure associated with the target.
>
> *Stage 2:* This stage increases the contact with the target. Information obtained in this stage includes colors, surface textures, temperatures, tastes, smells, and sounds.
>
> *Stage 3:* This stage involves an initial sketch of the target.
>
> *Stage 4:* Target contact in this stage is quite intimate. In Stage 4, the unconscious is allowed total control in "solving the problem" by permitting it to direct the flow of information.
>
> *Stage 5:* This stage obtains details regarding particular structures, such as the furniture in a room.
>
> *Stage 6:* In this stage, the remote viewer can conduct some guided explorations of the target. The viewer can engage some limited conscious

intellectual activity to direct the unconscious to do certain specified tasks. This is where timelines and geographic locational arrangements are analyzed. Advanced sketches are also drawn in this stage.

Stage 7: This stage is used to obtain identifying information relating to the site, such as the name of a location.

structure—the formal procedures of SRV. "Remaining in structure" refers to a viewer closely adhering to these procedures during a remote-viewing session.

target—something for which SRV is used to obtain information. Typical targets of SRV sessions are places, events, or people. More difficult targets can include a particular person's fantasies, the cause of an event, or even God.

target time—the date and time of the target, such as 1947 for the Roswell Incident.

REFERENCES

Andrews, George C. 1993. *Extra-Terrestrial Friends and Foes*. Lilburn, Georgia: Illuminet Press.

Fowler, Raymond E. 1990. *The Watchers: The Secret Design Behind UFO Abduction*. New York: Bantam Books.

Good, Timothy. 1987. *Above Top Secret*. London: Sidgwick & Jackson.

Hibbert, Christopher. 1982. *Africa Explored: Europeans in the Dark Continent, 1769–1889*. New York: Penguin Books.

Hopkins, Budd. 1987. *Intruders: The Incredible Visitations at Copley Woods*. New York: Ballantine Books.

Jacobs, David M. 1992. *Secret Life: Firsthand Documented Accounts of UFO Abductions*. New York: Simon & Schuster.

Mack, John. 1994. *Abduction: Human Encounters with Aliens*. New York: Charles Scribner's Sons.

Maharishi International University. 1990. *The Maharishi Effect*. Fairfield, Iowa: Maharishi International University Press.

Maharishi Mahesh Yogi. 1967. *Bhagavad-Gita: A New Translation and Commentary*. Washington, D.C.: Age of Enlightenment Press.

————. 1995. *The Science of Being and the Art of Living.* New York: Meridian.

Mavromatis, Andreas. 1987. *Hypnagogia: The Unique State of Consciousness Between Wakefulness and Sleep.* New York: Routledge.

McMoneagle, Joseph. 1993. *Mind Trek: Exploring Consciousness, Time, and Space Through Remote Viewing.* Norfolk, Virginia: Hampton Roads.

Monroe, Robert A. 1994. *Ultimate Journey.* New York: Doubleday.

————. 1985. *Far Journeys.* New York: Doubleday.

————. 1971. *Journeys Out of the Body.* New York: Doubleday.

Oates, Robert M., Jr. 1990. *Creating Heaven on Earth: The Mechanics of the Impossible.* Fairfield, Iowa: Heaven on Earth Publications.

Orme-Johnson, David W., Charles N. Alexander, John L. Davies, Howard M. Chandler, and Wallace E. Larimore. 1988. International Peace Project in the Middle East. *Journal of Conflict Resolution* 32:776–812.

Orme-Johnson, David W., and John T. Farrow. 1977. *Scientific Research on the Transcendental Meditation Program: Collected Papers,* Volumes I–V. Fairfield, Iowa: Maharishi International University Press.

Royal, Lyssa, and Keith Priest. 1992. *Visitors from Within.* Scottsdale, Arizona: Royal Priest Research Press.

Strieber, Whitley. 1995. *Breakthrough: The Next Step.* New York: HarperCollins.

————. 1988. *Transformation: The Breakthrough.* New York: Avon Books.

————. 1987. *Communion: A True Story.* New York: William Morrow and Company.

Swann, Ingo. 1991. *Everybody's Guide to Natural ESP: Unlocking the Extrasensory Power of Your Mind.* Los Angeles: Jeremy P. Tarcher, Inc.

Targ, R., and H. E. Puthoff. 1977. *Mind Reach.* New York: Delacorte Press/Eleanor Friede.

Wilber, Ken. 1977. *The Spectrum of Consciousness.* Wheaton, Illinois: The Theosophical Publishing House.

About the Author

Courtney Brown, Ph.D., is an Associate Professor of Political Science at Emory University in Atlanta, Georgia. His academic specializations include nonlinear mathematical modeling of social phenomena, environmental politics, democracy in developing societies, and elections. He held the Charles Grove Haines Professorship at the University of California, Los Angeles, and was a Hewlett Fellow at the Carter Presidential Center.

· A NOTE ON THE TYPE ·

The typeface used in this book is a version of Baskerville, originally designed by John Baskerville (1706–1775) and considered to be one of the first "transitional" typefaces between the
"old style" of the continental humanist printers and the "modern" style of the nineteenth century. With a determination
bordering on the eccentric to produce the finest possible printing, Baskerville set out at age forty-five and with no
previous experience to become a typefounder and printer (his
first fourteen letters took him two years). Besides the letter
forms, his innovations included an improved printing press,
smoother paper, and better inks, all of which made Baskerville
decidedly uncompetitive as a businessman. Franklin, Beau-
marchais, and Bodoni were among his admirers, but his typeface had to wait for the twentieth century to achieve its due.